INTERDIMENSIONAL PHYSICS

The Mind and The Universe

INTERDIMENSIONAL PHYSICS

The Mind and the Universe

by

Ruth Norman

Unarius Publications
145 S. Magnolia Avenue
El Cajon, CA. 92020

INTERDIMENSIONAL PHYSICS
The Mind and The Universe
A Unarius Book / March 1989

For information address: Unarius Educational Foundation.
145 S. Magnolia Avenue. El Cajon, CA. 92020-4522

Library of Congress Cataloging-in-Publications Data
Norman, Ruth
Interdimensional Physics, The Mind and the Universe

Includes index
1. Cosmology 2. Space and time.
3. Consciousness. I. Title
1989 89-90105
ISBN -0-935097-15-5

PRINTED IN THE UNITED STATES OF AMERICA

Dedication

The great help and assistance extended by the scientists from the higher dimensions, on the 'other side' of the universe, made this book a reality! These humanists are members of the Unarius Brotherhood, who individually and as a whole, have left traces of the logic and reason necessary to advance humanity to the next level of evolutionary progress. They number in the thousands and are the 'pioneers' who have laid down the basis for a physics and cosmology to reconnect man with the reality of energy as a universal plasma, interconnecting all factors of life within an evolutionary schematic.

To Ernest L. Norman and Nikola Tesla we owe special gratitude for the *Post holes* they have dug into this planet Earth, to provide mankind with the knowledge of the electromagnetic nature of the mind and the universe. They have given the *Keys of the Universe to Awakening Man*, to enable our planetary society to solve those problems that have been enigmas up to the present twentieth century. The result will be a *new age of logic and reason and a renaissance of Spirit!*

CONTENTS

Preface		i
Introduction		1
1.	Quantum Mechanics, Relativity and the New Cosmology	15
2.	On the Nature of Physical Reality	30
3.	The Infinite Concept of Cosmic Creation	42
4.	The Enigma of the Plasma Universe	56
5.	Atoms, Man and the Universe	76
6.	Energy and Matter, A Field Concept	94
7.	Parallel Universes - Matter & Antimatter	109
8.	Discussions On Quantum Solar Mechanics I	122
9.	Discussions On Quantum Solar Mechanics II	130
10.	From Quarks to Quasars	143
11.	The Sine Wave Principle of Life	162
12.	The Field Equation of Inner Space	180
13.	The Physics of the Higher Dimensions	202
14.	Regeneration versus Reductionism	220
15.	Vortexal Mechanics & the Venturi Tube	233
16.	The Non-Physical Nature of Man	251
17.	The DNA Connection	272
18.	Man, the Electronic Computer	289
19.	Towards A Physics of Consciousness	306
20.	The Renaissance of Logic and Reason	320
Index		333

Ruth Norman — URIEL
Cosmic Visionary

PREFACE

This book is different than other books that have been written on the subject matter of astrophysics, particle physics, mathematical physics, plasma physics, cosmology, electronic engineering, etc. Interdimensional Physics, the New Science of Life, unites the interconnected cellular structure of man's physical anatomy, his mental anatomy and the Universe. In this respect then, this book is an interdisciplinary approach to resolve once and for all, the enigma and dilemma of the separate nature of knowledge that has stratified the scientific community.

For the past hundred years, science has not progressed to a greater extent in the application of the principles of logic and reason. When mental constructs are used as a proposition to unite the mind, the body, and the brain as a trilateral expression of a higher intelligence, the higher equation of consciousness will reveal the fourth dimensional nature of the pulsing universe!

The history of man is a history of time and space. Time and space is the matrix of a yet unperceived plasma universe. The history of the terrestrial dimensions then, is a dim light that is, to all intents and purposes, yet to reveal the plasma from which our universe, in its immensity and grand design, is constructed.

The information contained within the chapters of this book, is laced with the imprint of advanced intelligences who have penetrated the cloud cover of the global community in which man lives on his Earth planet. They extend the frontiers of present knowledge to reveal the overall pattern of life as a latticework formed within the electromagnetic flux fields of energy, as a continuity of past millenniums of time, and of the future as an interdimensional progressive development.

The widespread and conflicting theories for the creation of the universe present a dilemma for present understandings of cosmology. In this respect the theory of the big bang for the creation of the universe, the red shift and Doppler effect, have interfered in the progressive evolution of man on planet Earth.

The schematic of life is interdimensional, with man as a protoplasmic energy system revolving in the immensity of the warp and woof of space/time, until he distinguishes the plasma of an infinite creative intelligence which is the substance and purpose of his evolution.

As science is the indelible ink that attempts to trace an image of the infinite universe, scientists should be able to shine light into the minds of man as to the real picture of the universe which will open a wider window of space, revealing the electromagnetic field forces that are the etched design of the plasma universe!

The reader will find in this book, the imprint of higher intelligences and the reality of the non-atomic factors of energy, light years beyond the terrestrial dimension. In other words, you will find in the chapters of this book, discussions of the various disciplines, not separated from one another, but relative to each other, revealing the creative factors of energy and of an interdimensional design of life for man. Thus, the reader will experience a transcendency beyond his physical consciousness as he will view his present environment, within the space/time continuum of an infinite universe.

Man is the theme of this book. Man is the unified equation that becomes the bright light, a consciousness, which will reveal the pattern of the 'Cosmic Plasma Universe'. Therefore, the paradigm of life must contain the supportive basis of all disciplines of knowledge. It is this free-flow of information, that provides knowledge of the principles of the oscillating universe. Then will life for man become more than a mirage. It will become a 'reality' that is substantive, based upon the individual's own inventory of experiential life, plus his interaction with the higher constructs of the plasma universe.

Interdimensional Physics describes the evolutionary principles of the plasma universe, which is the imprint of the pulse of creation, sustaining atoms, planets, solar systems, galaxies, universes and man, in a continuity of consciousness that is an infinite perspective.

In conclusion therefore, the authors of this book serve as a catalytic force for an advanced community of individuals who are members of different disciplines, but are functioning in a higher society beyond space and time. Such intelligences are the resolution of a higher imprint of logic and reason which when incepted, will remove the restrictions of the past that have impeded man's understanding of the electrodynamic processes of evolutionary life. Most important, is the individual's prerogative to advance the infinite imprint within himself, to become a polarizing Light to other men in their pursuance of a progressive evolution.

THE CASE OF MIND OVER ANTIMATTER

Paul Dirac (row 2, fifth from left) attended an international conference on physics in 1927, along with Max Planck (row 1, second from left) Madame Curie (row 1 third from left) Albert Einstein (row 1 fifth from left) and Niels Bohr (second row, far right).

Acknowledgements

We would like to express our gratitude for the editing and typography to Carol Robinson for her unselfish efforts in preparing this manuscript for publication; and to Dorothy Ellerman and Dorothy Mandragas for earlier manuscript preparation. Also included are Michael Leas and Ronald Robbins for color preparation and printing.

Thanks are also extended to Kevin Kennedy and David Reynolds for graphic diagrams.

For the use of the photograph of the solar corona, taken from data supplied by NASA's Solar Maximum Satellite, and the photograph of the sun taken by NASA's skylab, appreciation is extended to the Space Physics Division, NASA Headquarters, Audio-visual Branch.

Credit for the photograph of the sun taken by white light coronal camera is extended to the National Center for Atmospheric Research at the University of Colorado, Laboratory for Atmospheric Space Physics.

INTRODUCTION

"Michelangelo Buonarroti is noted as a sculptor, painter, scientist and architect. He is noted for the grand and great frescoes that are presently the inspiration for hundreds of millions of people of all religious affiliations and denominations. Michelangelo is known through his work which has inspired many hundreds of millions of people who have seen in the reflections of photographic slides and lithographs, the creative nature of his consciousness.

"However, the creative mind of Michelangelo has not been appreciated to the higher degree that it could have in the milieu in which he functioned in the 16th Century. Art portraying the history of man, both in the material sense and the spiritual sense, was commissioned by Popes and other religious leaders during the Reformation and Renaissance from the 14th to the 18th century. These notable works of art are presently the inspired motifs adorning the walls and ceilings of churches, art museums and galleries around the world. The Vatican contains the continuity of many Renaissance artists who were inspired from another dimension beyond space and time.

"Michelangelo and Leonardo da Vinci lived at the same time in Italy in the 16th and into the 17th century. Michelangelo's work had a dominating theme, life's continuity and meaning. Touching man with the theme of God, is the central scene of the main ceiling fresco on the great dome of St. Peter's Church in the Vatican. This is the universal theme of man's quest for his spiritual nature, the theme of God, the Infinite Creative Intelligence reaching down into the lower dimensions of terrestrial planets to complete the universal circuit of God and man. Of course, the entire theme is an allegory, as God is not and could not ever be indicated in a human form.

"The human form, the physical anatomy is the prenatal image for man to develop the conceptual understanding of the abstract nature of the Cosmic Mind, the Fountainhead, the Father, the omnipotent and omnipresent Force that is the all and all and from which all finite, third dimensional and higher dimensional factors of life are created.

"Leonardo, sculptor, painter, poet and scientist, developed the theme of man's struggle to understand himself throughout the many pressing conflicts involved in the battle between his lower and his higher nature.

"At the same time the paintings of Leonardo da Vinci indicate the great hostility at that time in the 16th century, This hostility between man and man was revealed on great frescoes portraying the battle scenes of war and the appearance of man's lower nature to indicate his negative consciousness. He indicated in this manner the profile of the earthman as he was yet a primitive aborigine who functioned no differently than many of the animals in their fight for security and the domination of beast against beast.

"Yet, as with Michelangelo, the portraiture of man as a spiritual being was his essential purpose. It was the inner direction of both artists, to indicate within the marble of the sculpture of either the horse or man, the essential ingredients which illuminated these sculptures in the pure flowing lines of the horse, in the spark that was shown to reside within the features of man, particularly in the eyes.

"El Greco, painting under the patronage of the church fathers was under great stress; yet was dominated by Spirit.

With the recognition and truth that existed within him, he revealed the essence of man, essentially as a flowing river of energy, containing the evidence of a science of life.

"Art, music, paintings, sculpture, and theater are representations of the Infinite. Manuscripts and books, carry the signature of the higher developed and more creative factors, a development within the mental natures of those individuals who occupy a higher position in a world of higher energy structures, containing the all-encompassing nature and attribute of the Cosmic Mind.

"The river of life flows in two directions. As man is picking up his palette, or his pen, donning the robes in his portrayal of some scene on the stage or raising his violin to his shoulder, he indicates the flow of this river of life which contains the greater intelligence of individuals who have developed the essential reality of Infinite Intelligence.

"Whosoever should draw his bow over the strings of his violin, run his fingers down the keys of a piano, raise his or her voice in an aria on the operatic stage, lift his hand and dip his pen in the inkwell to write a moving statement of the nature of the mysteries of life, are individuals who are receiving the note of Creative Intelligence from those individuals living in higher dimensions. Such persons, who themselves have functioned in these various themes of terrestrial life are now participating in the act of creation which is their inspirational purpose and sole objective.

"Leonardo da Vinci, El Greco, Michelangelo, Raphael and a host of other Renaissance and post-Renaissance painters, many unknown to the present generation, have inscribed upon the temple walls in Egypt, Greece, Rome, Atlantis and Lemuria, and on the walls of great standing buildings in other planets, the flowing river of events that have transpired on their worlds and in their civilization. All great works of art, in whatever medium, contain the continuity of intelligence, the continuity of the River of Life. They essentially speak from the higher dimension from which such inspiration has been created into the medium of the atomic form on some earth world.

"The progressive understanding then, of man as an evolutionary development, is the continuity of the conciousness of the greater intelligence of individuals who, although invisible from man's physical senses become visible to these artists in the mediacy of their creative inspiration.

"So it was with Michelangelo and with Leonardo in their vast portrayal of man as a continuity of consciousness, representative of the Infinite Creative Intelligence and indicating that the finite man or woman is not marked by the measurements of the height of his physical anatomy, the respiration of his pulse, the regularity of his heart beat, or by the encephalograph that measures his brain wave!

"Man is greater than all of the emissions which are presently discernible on the electronic devices; instrumentations that are carved from the knowledge of third dimensional life. Man, as he is a spiritual being, can be the true intent, the representations of great works of inspiration which have been rectified within the atomic picture of life. They are, in every respect, representation of spiritual life, in music, containing its imagery in sound, in painting, containing its imagery on the canvas, and in books, containing its imagery in the ink of the paper; also the singer and/or the actor containing the imagery in costume and voice. These are representations within the material element of atomic forms.

"But since the reflections of imagery have been received

from a higher sense of sight, and demodulated into the atomic world, they contain the psychokinetical frequencies of energy which touch the higher nature of man, his higher sense structures which are the paranormal factors of his real self!

"These higher sense structures called the sixth, seventh and higher senses are the basic spark that is being fashioned and amplified through inspirational measures which represent to all mankind the recognition that there is life beyond the physical material dimensions of his present atomic world. They represent the higher mind which contains a structure that is not dependent upon the five senses for its life and breath.

"Man contains a structure which is, to all extents, a higher amplification system, an electromagnetic device which must needs be learned and made familiar. He must learn to play the chord structures of this electronic device. It is the higher organ that allows him, as the artist, to become a channel so that he may receive new information. Through this channel will flow the identities of people who now express infinity on higher worlds and who have fashioned their own instrument to receive the harmony of the higher dimensions. The notes of the piano, the strings of the violin, the sounds from the wooden instruments have been expanded to receive information of the reality of man as an imperishable factor of higher Intelligence.

"Man is thus learning in all stages of life through different mediums, the nature of himself. He is learning that whatever this medium may be, it is a medium that validates the science of life; that life is an energy system, and that this energy system, infinite in its creation, contains an infinite number of frequencies.

"The instrument for man's physical life consists of the five elementary physical senses which fashion his ability to receive information from the atomic environment of his planet. The number of keys that are the instrument panel of the Infinite are themselves infinite and man is just beginning to operate the lower keys of this electronic device.

"Essentially, man is an electronic device formed from an infinite electronic system. He is beginning to learn the principles by which higher frequency devices function. The essential nature of life, is the difference contained within the frequencies of the various keys of the piano or the various strings of his musical instrument. Man is playing upon his own instrument in this fashion and he is learning about the basic frequency of himself, while he lives within the atomic arena of life. He is learning therefore, to recognize the sensitivity of his five-sense body which is a lower frequency structure.

"The higher frequency which is an indication of the non-atomic nature of man, is contained in the medium of the creative work of the artist, the musician, the actor, the writer, the singer, and those who are expressing the pure abstract concepts of life in philosophy and science. These individuals are beginning to reveal in the 20th century, the essential identity of the sixth and the higher seventh sense structures that are presently the next development of man.

"At this junction in epochal time, the river is flowing to a higher crescendo. The movement of the *River of Life* contains the basic history of man from his five-sense structure of a physical nature, and his sixth, seventh and higher structures, and is the evidence of the reality of his own true energy system. It is this musical score which each individual has been instrumenting in the long history of mankind as he was learning to play the keys of his own instrument. Many of these keys have been out of tune with the essential nature of the design of his instrument and hence, he has been functioning essentially to correct the malfunction of these keys which is the nature of his recognition of the feedback that this information has given him.

"The corrective means by which these changes are being made is the basic development that is now taking place. It is taking place with respect to the reality that has been brought to his higher sense structures in the flow of life and the continuity of consciousness. This is the intelligence of each individual who has developed to a point where he has contained the more corrective and positive factors of the energy structures of Infinite Intelligence. Essentially, man is learning then that he is not the rude, primitive envelope which he feels is purely his physical anatomy. He is not simply flesh, blood, bone, and other organs that constitute the physical anatomy.

"In some respects man lives through the atomic cell structure until that time when he can penetrate through the atomic frequency of the instrument he occupies and is able to distinguish the higher musical tone that is his true nature. It is this higher musical tone which is the frequency of those individuals who have been alerted to their spiritual senses; a body which functions beyond the frequency of the five physical senses and is a factor of and a 'wholly owned subsidiary' of the infinite creative intelligence.

"In this respect we can indicate that man is a subsidiary of his higher self! This subsidiary is the physical anatomy. It is the prerogative of man and his purpose to make apparent to him the reality of the higher energy body which as a river, is constantly activating and polarizing this atomic image.

"The motivating force of the psychic anatomy that is constantly expelling its intelligence into the cellular formations of the physical anatomy is the chief body, the controlling body. The physical anatomy is but the subsidiary, receiving whatever has been already impacted and developed by this instrument that contains within it the intelligence of the moving Force, called Infinite Intelligence.

"Man is now reaching that point in time in his third dimensional life wherein the great cyclic formation of this Infinite Intelligence is now rendering itself known in the amassment of the finite developments, which is the history of mankind on the planets of galactic systems. The immeasurable knowledge of the finite factors of life, is itself proof of a greater force which is constantly moving and ever changing, containing within itself the revealing characteristics of its nature. The limiting nature of third dimensional research is the evidence of its inability to explain the continuity factors of life. It is just this enigma that has been a dilemma and mystery to the man of science who functions basically in a compliant manner in carrying out the basic designs of his masters.

"Individuals who have touched a higher awareness of the electronic nature of their infinite instrument have revealed through the channel of their own consciousness the evidence of the higher keyboard which is contained in their own electronic device. This higher keyboard has alerted them to the existence of advanced human beings who can function beyond and above the lower five senses. To this extent each of those individuals who are represented in the archives of museums and art galleries, libraries, theater and science reveal the true nature of man as a reality of infinite proportion.

"The very same forces that inspired Michelangelo, Leonardo, El Greco, Raphael, and many others that cannot be mentioned in the space of this book are the very same Intelligences who have activated the minds of man in previous lifetimes and civilizations. These master artists have not only functioned from the fourth and higher dimensions, an abstract perspective for physical man, in order to integrate the continuity of the life principle, but they have taken on a material body, and through the principles of their own creative knowledge of the Infinite, materialized into a physical world.

"They have taken up the palette, the chisel, the baton, and have rendered the great arias on the stages of the opera houses, and in the symphony orchestras. They have taken pen to paper and written of the truth through the images that have been carried upon the pages of the Books of Life as a progressive development and of the reality of man who can live removed from the primitive, elemental, physical anatomy.

"This is man's destiny! This is man's evolutionary birthright and it is the nature of man's own nature to evolve from a caterpillar to a butterfly, to fly free of the branches of one tree to move about and revel in the beauty of the landscape that opens up to him as he is able to sip the nectar of the bountiful flowers in different meadows; to be nurtured in this fashion to again change his form to even another feature of Infinite Intelligence.

"Man can therefore be seen to be an essential spark of the Infinite, capable of moving up the ladder of evolution; to enter into new states of consciousness with the higher sense structures that he has become alerted to scientifically. He will then play upon his electronic instrument created from and by Infinite Intelligence, and be able to view the wider landscape of worlds which are non-atomic, providing him with the overture of a grand and great symphony, one that becomes the conscious participation of such an individual.

"Man, in his finite form, cannot be considered an individual until he can appreciate the nature of his individuality containing the basic structure of the creative energy of the Infinite and can play upon his electronic instrument. Then he begins to realize and understand that he is a member of an orchestra and must be in tune with and participate with the orchestra.

"It is this knowledge that makes man more than one instrument and capable of existing in the greater awareness of the medley which is the harmonic overture in which he becomes a part. Thus he expands the nature of his own infinite nature by his participation as an instrument of this orchestra which is the accompaniment of like souls who have learned to play their instruments, each one different than the other, each giving off a different frequency.

"Thus it is the sum and total of the higher frequencies that make up what has been called in literature of man's past, the music of the spheres or heavenly music. It is heavenly music, because it is beyond the gyrating lower frequencies of the physical senses containing within it the chorus of the energy of individuals who have blended their own frequency with the frequencies of the hundreds of trillions of individuals who are now so integrated that they make up a unified consciousness. And it should be each individual's desire to become a part of this whole. Man must learn then, the discipline of playing upon his instrument. He must have first-hand knowledge of this instrument that has been crafted by a greater Intelligence and he must then learn to play upon the keyboard of this instrument.

"Man is an instrument of infinite proportions and he contains within it unimaginable possibilities to extend the awareness and perspective of his consciousness! He therefore begins in the atomic body through which the lower keys are learned and developed to the extent that he gains certain appreciation for the complexity that is the nature of the physical anatomy, those sense structures which are the keyboard that provides the acknowledgement of its movement.

"Yet this body is only the subsidiary instrument. It is the student's harp. It is the child's piano. The true instrument now must be learned, an instrument which provides the force and the greater intelligence and which itself can activate and develop a newer instrument, updated and renovated where it can function in an environment that will be free of dust and corrosion.

"This is the present development of mankind in the earth worlds spread throughout the billions of galaxies in the universe! This is man's beginning within the Milky Way galaxy. He is presently learning about the primary instrument which is his electromagnetic body - his mind. He is beginning to learn of the existence of such a primary energy system and the learning must be gained by lessons which become the basis of his new education.

"This is the essence of our discussion with you at this time, to reveal to mankind that he is about to step out of the cradle, having learned of the primary nature of himself and the ability to function within the limited scope of this secondary body.

"This is the grand and constant flow of the river in its sinusoidal expression, moving in its proportional manner from the nucleus of its centripetal field from which it had fashioned the atomic worlds of the terrestrial dimensions and had breathed life into those physical anatomies where man first gains the perspective of himself as a physical organism. In the understanding of the higher physics of his energy body, the electromagnetic body that is the true basis of all his sense structures, he must now begin to walk with the strength gained from the understanding of the movement of energy.

"The movement of energy is the nature of Infinite Intelligence. The movement of energy is the true physics of man's existence, the understanding that he is not a solidified mass that can only exist in one dimension. In this recognition that has been extended throughout the millions of billions of years in the nuclei-state of the beginnings of man in the recognition of his infinite proportion, the grand sweep of the Infinite is manifesting at the close of the major cycle of the Sixth Recessional. Advanced, intelligent human beings in non-atomic worlds are, as has been indicated by Michelangelo's Fresco, reaching their hand down towards man in his lower position and, in every essence, giving him a lift up. This *lift up* is the bilateral hand of higher intelligent beings who have inspired man through these many creative mediums in which life has been portrayed as an essential movement of life.

"Man himself, must also develop the platform from which to fly, to free himself with knowledge of the physics of his own mind. This knowledge will give him his first initiation into the heady aroma of a life that is not subverted by the lower senses and by the carnal appetites of the physical anatomy. This is the beginning of a new era, the spiritualization of material earth worlds, the spiritualization of those individuals who have prepared themselves and who are prepared to take up their bed and walk, having accepted responsibility for their actions in past life endeavors, which have essentially carved their own prison and have kept them in bondage to their lower senses!

"The higher notes of the keyboard that exists within each individual's mind is being sensed by more and more people throughout the earth worlds. And it is in this manner that the chorus is beginning to engage itself whereby the harmonic regeneration of this knowledge which is existent within the chorus and is the freedom of those individuals who have learned of their own inner dimensional and interplanetary function. As a stone rolls down the mountainside and gathers the force of its own momentum it finally comes to a stop, only when it has reached a point in this descending movement, where a force greater than itself, has halted the stone.

"This has been the nature of the degenerating movement of energy through the civilizations of mankind a movement which has been brought to a halt in its momentum through man's progressive awareness of the higher keys that exist in the instrument of man. The release of this energy that is the creative factor of Infinite Intelligence is now the spiritual movement of man as he is essentially reversing his direction, to add to this progressive movement which is the nature of spiritual understanding gained through the knowledge of interdimensional physics, the science through which he can create a great force.

"This force will be his arc and the trajectory, and as an arrow, will move according to the force that is expended by the individual through his mental endeavors. It will be in the understanding of the psychokinesis of energy that he will learn to always point his arrow so that it will fly in a higher trajectory and carry with it the force of his positive endeavors.

"This will be the essential difference between man's negative past where he has stopped the flow of the ingredients that would nurture the higher senses and would activate the keyboard of his mind. He will learn of the mechanics of flight and become a part of the new generation which will activate a channel from and through which the creative measure of life lived in the higher dimensions can flow into the third dimension, to give it the stimulating force which will develop the momentum for a progressive evolution. In this fashion we see the ever-cyclic movement of infinity, as it gathers back into its fold, the nature of its own reflections, drawing back within itself all of the characteristics which have been its basic ingredients and regenerating again the infinite in the outflow of its creative force.

"The breathing in and the breathing out is therefore an act of consciousness which is the basis of the understanding of this advanced science of life. Knowledge of interdimensional physics and its application will provide the momentum for a progressive development, whereby the individual will see the reality of himself as a unified field system, functioning in-tune with the infinite!

Chapter 1

QUANTUM MECHANICS, RELATIVITY AND THE NEW COSMOLOGY

"The consequences of Max Planck's psychic awareness which has become the present evolutionary design called Quantum Physics or the principle of the quantum, as a basic understanding of the movement of energy, extrapolated upon and within the language of third dimensional physics, has essentially rocked the establishment of the scientific community. The repercussions of this great giant leap forward into the understanding of the reality of energy as an interdimensional continuity, explains the unified state of consciousness that exists within the most minute factors of energy! This is true! The atom and the subatomic particles are unified states of consciousness. It is in this unified state that they exist and are related to a whole number, which is the nature of the expanded energy system from which they were formed as divisible, integral numbers or integer units.

"As such they are, in mathematical equivalents, truly and essentially the proof of the indivisibility of the Infinite and yet, are the divisibility of the Infinite! This is the basic proposition that has stymied mathematical physicists in their attempt to work within the boundary lines of a third dimensional language system, in the attempt to paint a picture of the inner reality of energy.

"Quantum mechanics has essentially done this and has irreducibly explained the reality of the microcosm and the microcosmic world; in essence, revealing an infinite intelligence functioning beyond the framework of man's five senses. The five-sense structure then, which is the interrelationship within and by which the framework of societies function, is a contrived means to contain some security within the overall existence of the macrocosm of the greater Infinite.

"It has been discovered that energy moving on sub-atomic levels is inconsistent with the large view that has been developed through mathematical physics. There is an order which is directly inconsistent and a dichotomy with Newtonian physics, that had previously held the universe to be a recognizable continuity of the basic laws in which physical planets are held in their orbits around their suns. Physicists have recently discovered that the interior of the atom is a vast unknowable sea of energy that functions without dependence upon the basic laws of momentum, inertia and gravity. This is a great enigma which is presently being recognized thus altering the overlaying structure of classical third dimensional physics!

"Physicists are beginning to become aware that the atom is not dependent upon the overlay that has been the structure of third dimensional physics with respect to the fact that there is a knowable boundary line within which life had developed. This is the understandable basis of the present theory of the initial beginning of life, the 'big bang' theory. Quantum physics violates this understanding! The theory of the quantum can therefore be seen to be the first visible evidence of the proof of the continuity of energy from the fourth dimensional perspective and that a new mathematics must therefore be designed.

"Robert Oppenheimer has had close association with the energy force that is discharged from an atom. He had long ago stated that man or science will need a new mathematics to explain the properties of the atom. Albert Einstein had extended this thought previous to Robert Oppenheimer and stated that energy is an intelligence of the whole number; that the whole number contains the factor of all its parts which is the real integer. The integer is therefore the basic knowledge that is being pursued by man and experimented by scientists.

"This is the basic feature of the third dimension - a dimension bounded by the speed of light. It is the radiant force of light as the basic whole number, which is the juxtaposition of physicists to determine all other factors that lie within this boundary line and which are the integers.

"The basic needs, therefore, which are the maps that have been developed and depended upon, have been within the nature of this third dimensional boundary line. Since any association of life beyond that boundary line would be inconsistent with present knowledge, life therefore, if there is such in their indispositions, would have to be life that could be evidenced by and recognized through the understanding of the structural field forces which are the principles of interdimensional physics.

"Einstein had been involved throughout his entire life in the quandary of the quantum mechanical theory of atomic energy, a force unexplainable by the classical laws of Newtonian physics. Bohr also realized from another perspective, with his theory of complimentarism, that the ground rules of physics could not provide a deeper understanding, interdimensionally of the particle nature of the atom, as this Force originated beyond the boundary of 'time and space'.

"Wigner and others are of a coterie of more advanced scientists who contain a higher awareness of third dimensional physics. These physicists are particular individuals who know that the boundary lines which are the exterior lines of the present third dimension, are not necessarily boundary lines but an indication of the continuity of energy as it is, in its consistent movement, constantly regenerating and changing its form. To all intents and purposes, this is the accepted feature of life lived within the boundaries of the atomic dimension.

"Before defining the dimensional quadrants containing the continuity of intelligence of man with new factors within the embracing fullness of Infinite Intelligence, an individual would already have to have within himself a development which contained experiential factors of his own association in higher fields of this energy structure. He, himself, would have brought back with him this awareness, which would be the means for the advancement of present knowledge of physical reality.

"Therefore, the very maintenance of the mathematical symbols as a means of proving the existence of life beyond the pale; that is to say, beyond the stockade, beyond the wall which provide the present security to these individuals, cannot be determined as long as he functions within a rudimentary symbolism. Symbolism, after all, can be advanced to an extent that it becomes not the symbol but refers to certain factors within the movement of energy. It is all determinant, of course, upon the ability of any individual to conceive a newer concept.

"In symbolism is the picture of a whirlwind. The whirlwind is highly impacted at the center or nucleus. This impaction results from the electromagnetic lines of force moving in a centripetal fashion into the nucleus, causing the great force of a tornado. Extending that symbolism to man in his evolutionary development, in the evolutionary movement of atoms, of planets, and suns, is the individual's ability to conceive of the underlying factors that make possible the presentation of a picture of life.

"The picture, therefore, can never be any greater or lesser than the factors that have already been developed within the mind of an individual. In every respect, the mind is the holding device, the computer which has already had programmed into it certain information accumulated in the development of that individual through countless life experiences.

"The essential understanding is the recognition of evolutionary changes coming upon mankind and discharging the obsolete information from man's past. A boat, to survive the rapids which it is attempting to move through, must lighten its contents, to be able of ride the rapids. In every extent, the rapids are symbolic of man's rocky road which has created the heavy boulders containing the verbiage and out-of-sync information based on the principles of third dimensional physics. Man's consciousness is the movement of his mind through life-to-life experiences, where man has been a basic traveler throughout the cosmos and has lived in a myriad of planetary systems in his evolutionary development!

"The river is man's life-line moving as a sine wave. He finds that he has been carrying with him material features of past life experiences that have become so compacted that they have developed into mental boulders. From one life to the next, he has thus been capsized in his movement through life, evolutionary-wise. In each capsizing he has regenerated the anger, frustration, insecurity, etc., in his attempt to reenter his capsized boat, which represents his security.

"That is well and good up to a point in the evolution of mankind; but when the river begins to turn and carries his boat into an unknown area, then he no longer has the means by which to navigate, as all factors of life have lost their relationship to him.

"This is the nature of the present development that is, in a greater way, taking place with the scientific community. Those scientists who have made that turn in the great river of life, have navigated into an unknown area. Others, in a fearful manner have remained within the boundary lines of the river which now represents their present life, protected with even thicker walls to reinforce their previous belief system as a means to garner their insecurity.

"There are scientists who are being carried along the river on the boat, which is their planet, and who as passengers, have no alternative but to accept the changes that are being shown to them in the environment, because the reality of this river is that it is a tributary of a greater river taking the passengers on their planet to a wide open sea. This open sea is the unknown to those individuals who have been living in the backwashes of the tributaries of the river!

"To all intents and purposes, the symbologies we present to you are the many tributaries which have been the beginning for man's evolutionary development on the third dimensional planets. But these planets are moving in a sinusoidal manner into an expanded factor of the interdimensional vortex, where life was first gestated. Man is an essential passenger on the boat, traveling through life's river revealing the expanded nature of life as he is journeying towards the wider opening of the river, revealing the existence of an infinite sea. This infinite sea is being revealed to man through the research of the scientific community evidenced with particle physics and cosmology.

"The interior of the atom is indicating to the particle physicists that there is no completion, there is no ending. As far and as deep as they probe they are becoming aware that in their probing they are finding the evidence of a greater energy force! Hence, it can be seen that the individual minute particles which make up the property and the proposition of the atom are as the tributaries of the river. Physicists are moving into the awareness that these tributaries had their basis from a wider sea.

"They are beginning to become aware that the interior of the atom is revealing the nature of the continuity of the force that had been recognized by Robert Oppenheimer as being a fourth dimensional factor of the greater Infinite. It was his great concern and his great sadness, to this extent, that this unknown force was allowed to be exploded unintelligently into the third dimensional field in which man lives, bounded by the properties of the light spectrum.

"There is a mathematics that explains the physics of interdimensional energy. But this higher mathematics must be a development within the mind of certain scientists who will be responsible for initiating this advanced mathematics. The interdimensional physics presently known as Unarius is the basis and the foundation for such a mathematics. But as long as man functions within the properties of the nuclear form of energy, his symbolism will be a functioning part of the light barrier. To work with principles of energy that are non-atomic would clearly indicate the existence of life in nonatomic worlds. Such an individual would no longer be a third dimensional being; he would be a fourth dimensional being. And yet there is a means by which the representation of the fourth dimension can be brought into third dimensional understanding.

"The New Physics must be an interrelationship between the physics of Copernicus, Kepler and Newton, and the pioneering theories developed by Planck, Einstein, Schrodinger, Bohm, Maxwell, and others of equal renown.

"The symbolism of calculus will have to be replaced so that man can successfully navigate the rapids of obsolete factors of the nature of physical reality! He has allowed the obsolete measurements of his universe to take a greater hold on him; obsolete concepts which have been strengthened because of childlike dependence upon his five-sense body.

"Man on all fronts is beginning to develop the logic and reason to conceive that he alone contains the answers to the many propositions for the explanations of the enigma of life. It cannot come from the continuity of past methods, using only the square and the ruler! He must begin to work within the proposition of the circle. He must begin to conceive the fourth dimensional vortical nature of energy to remove the obstacles of ignorance. This ignorance is a basic boulder formed from the momentum of worn out dogmas!

"No scientist in his research for the explanation of the physics of energy can further this development without the realization of his own past life experiences. In his present acceptance or unacceptance of the newer physics, he is, in all respects, functioning from some association of events which have caused him his present indisposition and inability to further his own mental constructs as an energy system functioning within a larger energy structure, i.e., an electron, moving within the greater electromagnetic force field of the atom. Therefore, he has not recognized his association with the protonic forces of life. It is this disability that causes the dichotomy within the mind of such hard-nosed scientists who are incapable of dissociating themselves from the dogma of a one life condition.

"Science in this respect can be seen to be not pure, in

that the elements which are the accepted objective factors of the reality of energy are also an association of the subjective energy force of the individual. Therefore, in every respect, there is always the objective exterior force which is the finite movement of whatever structure is being researched. And there is the researcher, the individual who subjectively is attached to previous dispensations of knowledge which is the continuity of his past concepts.

"There is an ever-widening need to realize the very essential factors of energy as an ever-moving, expanding force, constantly changing the nature of its force field in such movement. And when the energy is revealed beyond the expected latitudes of research, then the mathematical factors of the symbology must give way to a new reality, and a greater understanding and recognition that the Infinite, in the finite properties of itself, contains the information of an infinite concept of cosmic creation.

"Planck's Constant, states essentially that energy functions in a different relationship within the interior of the atom than from the exterior of the atom. That is to say that the essential nature of life as the accepted mode of existence of mankind, which is the atomic feature of his environment, his physical body, the buildings of a city, the features of his material environment - mineral, vegetable, plant, animal and man - are only an apparent reflection of life. But essentially, in the continuity of life which is seen to be within the inner form of each of these exterior factors of life, there is a fourth dimensional movement where energy is functioning with a higher frequency and a greater force, and in all respects is the interdimensional proposition for life.

"This is the new development which is beginning to break down certain obstacles that have been the conceptions of the nature of physical life with respect to the indivisible movement of energy. To continue the voyage over the rapids of third dimensional life concepts, will harden the dogma of obsolete information. Continuing with such faulty navigation would surely wreck the boat should he not change his direction and cast overboard those factors of life which have been associated with out-worn information. Making the turn will indicate another development and the appearance of life on a broader scope, and the opening of man to a functional relationship with the Infinite as a whole number. Man was always an integer really, not a fraction of the Infinite, but at all times has been, is, and will always be the expression of infinite intelligence.

"It is this basic barricade that is the difficulty of most third dimensional scientists who do not know of the existence of the wholeness of themselves which has caused them to look for answers to the enigma of the grand proposition of Infinity from exterior forces within the third dimensional environment. The very structure and function of the atom lies within each individual who is no less than his own expression of his past-life experiences and his association in life-to-life expressions on terrestrial planets.

"The basic proposition of evolution is that man advances with the full help of individuals who have developed to a greater degree the fourth dimensional physics of life in their life-to-life development. With the cognizance of intelligent life in the expanded universe beyond the nucleus of third dimensional worlds, one receives in the self-same manner a greater intelligence with his ability to communicate and receive newer information which strengthens his awareness of the many worlds in which man lives to express the Infinite.

"Man replicates the Infinite and recreates a new environment containing energy of a higher frequency. With great stimulating relationships, man reflects from the frequency of another environment - cities of crystal that oscillate at higher frequencies than do the basic 109 elements of his present atomic structures.

"Man, therefore, is at the present time, entering into a new age which is being reflected throughout the works of more intelligent individuals who are aware of the infinite perspective of life. These are the pioneers, who are shoring up society for a new civilization, not bounded by the physics of a third dimensional proposition of life, but will contain the awareness of the continuity of man as a fourth dimensional energy system.

"We can also state that Fermat's theory of whole numbers and integrals can be answered, not from the mainstay of the mathematical calculus but must be answered only within the proposition of the physics of man as an indispensable reality of energy and not as a theorem. The third dimension, in all respects, can therefore be considered in this symbology to be an integral of the fourth dimension and the fourth dimension to be an integral of the fifth, which will be an integral of the sixth, etc. Therefore, in this awareness, one can accept the proposition that these so-called numbers are everchanging. In this change can be seen the validity of infinity, because the infinite can never be bounded, but is a regenerative oscillation. Its integrals can always be associated as participants or as finite resolutions of his own fullness or wholeness.

"The only manner in which any scientist can further his development is with the realization that the Infinite is infinite; in that, it is ever regenerative; it is ever moving; its movement is a whole number. This whole number is whole, in that no matter what its denominator, it will always be larger than the numerator. Therefore, the integer will contain the nature of that intelligence which was abstracted from the whole number.

"The proposition, therefore, of an expanding universe is central to the understanding and to the acceptance of any mathematics which would attempt to determine the present feature of life in any dimension. However, in the third dimension the proposition, and the understanding of energy, is more difficult, in that the individual who is subjectively involved has not yet attained to that understanding of the whole number; that is to say, he has not attained to his or her awareness of his own infinite proposition. He has not yet accepted the innate factor of his relationship to an infinite creative intelligence. And although an integer, he views himself subjectively and separate from the denominator. He cannot see the interrelationship between himself as an intricate factor of the infinite, to the whole number, and yet cannot exist without this whole number which number is the essence of his own higher self or superconscious mind.

"This leads us directly back to the basic statement that a scientist can function within a specific integer or that is to say, within the basic dimension in which life is the evolutionary statement for the progressive evolution of any factor of life. Man, as Homo sapiens, is the finite factor of life. The individual is thus developing through his own experiences, life to life, certain knowledge of the intricate wholeness of his mentality, which will regenerate the infinite nature of creative intelligence..

"Yet there are scientists who are still functioning within the parameters of obsolete mathematical symbols.

However, there are fourth dimensional physicists who are aware of the fact that newer relationships must be developed describing a consistent explanation for the interdimensional nature of Infinity. Individuals like Wheeler and Bohm are examples of scientists who are developing a mathematics to describe the fourth dimensional nature of the energy, validating the thesis that subatomic energy does not coincide with the movement of macrocosmic field forces within the third dimension.

"To all intents and purposes, man is becoming aware of the fourth dimensional factor of life as it is evidenced in the deep space of the nucleus of the atom, which is the protonic force reflecting the fourth dimensional pressure, or field, of higher dimensions.

"The infinite cannot be known except through its finite features. The finite reveals in this manner the infinite, and explains the great dichotomy of life functioning within each finite particle. But since the finite is an endless feature of life in the third dimension and is an endless feature of life in the fourth dimension and in all dimensions, this is indicative of infinity. Man, as an energy factor, proves to himself that there is no beginning or no ending! Life therefore reflects the immensity of cyclic movements, of wheels within wheels, that the smallest wheel within the great macrocosm of infinite intelligence is the atom.

"The atom can therefore seem to be finite and yet infinite. In its finiteness, it expresses the infinite because it is not singular. There are no reflections presently conceived by man through his physical senses or indeed within his higher psychic senses, that are separate from the great whirlpool which is the vortical movement of energy, ever widening in its expansion, ever decreasing in its contraction.

"Man in his relationship to the whole number, is always the evidence of the infinite. There is truly no separation. No matter where you place your finger you are cutting through the electromagnetic lines of force which join the energy fields of the most minute micro-organisms with the macrocosmic forces that make up the sum and substance of planets, suns, galaxies, and universes.

"Atoms, man, planets, suns, galaxies, therefore, are all evidence of the nature and interconnection of life. There is at all times, a constant transposition from the transmitter to the receiver. Hence the great enigma facing man, is the nature of extraterrestrial intelligence, the reality of Homo sapiens living on other terrestrial worlds within the galactic system of the universe.

"The answers to all insoluble mysteries are within each individual's mind. The mind, therefore, is revealing the picture of the enigmatic movement of energy containing within its own field force the telltale design of the interdimensional movement of energy and with it an interdimensional understanding of science.

"Science, therefore, can no longer be called science if it does not take into consideration interdimensional principles of energy, and how each individual contains the proposition of his own spiritual identity as an interdimensional energy system. The voice that speaks contains the fingerprint of any individual. Mentally and physically he is an interdimensional structure of relationships containing the history of himself from his very lengthy past of millions of years and containing also the nature of his creation. Thereby, we find that it is in the understanding of man that the understanding of science will be furthered.

"The revolution of quantum physics is thereby validated

and advanced with the understanding of the quantum mechanics of Max Planck, Erwin Schrodinger, Werner Heisenberg, Albert Einstein, Dirac, Pauli, etc. Hence, these individuals are the proof of the continuity of the interdimensional understanding of life. These are individuals who carry with them the potent powers of an expanded understanding of life and reiterate this truth by demodulating such understanding into the language of man. Society is a continuous development of progressive features by individuals who are transmitters from a higher dimension, carrying in their transmittal tapes the knowledge of the existence of fourth dimensional concepts - a fourth dimensional physics, if you will!

"This is indeed proof that man is on the road, or shall we say, he is moving through the river of life and is beginning to take a new direction into a new development. As he turns around the bend he is, becoming witness to a higher refraction of the electromagnetic spectrum where he will find the validity for his own infinite nature. Therefore, in this respect he will become a brother and will be incapable of raising his fist to another as he will be smiting himself in that gesture.

"The revolution, in physics is an indication of the evolution of energy in its constant cyclic interplay, regenerating through electromagnetic lines of force other factors of the infinite. It is in this respect that advanced intelligences are living in higher frequency worlds of the infinite, where the whole number is their integrated intelligence. Such individuals are no longer separated by the atomic nature of life but have attained a higher perspective, enabling them to see beyond the boundary line of only one dimension!"

Chapter 2

ON THE NATURE OF PHYSICAL REALITY

"A coterie of advanced people are aware of changes presently a factor of research into the nature of the mind. Individual scientists have emerged from the subsurface, so to say, of theoretical considerations of the enigma of third dimensional life to advance the expression of man's fourth dimensional quest for an understanding of self. In the scientific disciplines, such as biology, geology, paleontology, astronomy, physics and cosmology, etc., there is evidence that man is slowly progressing in his knowledge of evolutionary physics, because of the interdimensional factors associated with the phenomena of physical life.

"There has been a growing acceptance within the scientific community of former secretive avenues of thought labeled metaphysical, or occult. This development is resulting in a freer expression and interest into the matrix of the mind, advancing new avenues of study of man's evolutionary design.

"It is well to recognize that Johannes Kepler, a scientist of great repute, was looked at askance in the many facets in which he entertained a concept of life which included other terrestrial planets, as a factor of a higher reflection of Intelligence. Newton, the recognized 'rock of third dimensional physics', was himself heavily immersed in what at that time in the 17th century, was called occult. The evidence of their great works explaining the physics of energy, was a result of their inquiry into the invisible cosmos which they had been able to rectify into the language of the mathematics and physics.

"Descartes, another scientist of great repute, considered to be the Father of modern philosophy physiology and psychology as well as analytical geometry, set down the foundation for the inquiry into the nature of the mind: "I think therefore I am." The development of science, into new areas of the vast sea of infinity has been extended through the works of those particular scientists who carried with them the 'Whispers' of eternity and were able to formulate these whispers into a language which could be accepted and could be reflected out into the life of man as he exists within the atomic boundary line of a lesser infinite.

"Wigner himself has given proof that science is not a relationship of symbols only within the framework of existing concepts gathered basically from the evidence of previous analysis that had provided a logic and reason for the present understanding of life. We are speaking of life as it is seen by scientists, relationships of many unseen factors which cannot be validated by instrumentation. Yet it is just on this basis that all scientists agree, that the researcher is no more than his instrument; that there is a subjective element in all objective inquiries into the nature of life. Nature is being recognized to a greater extent as an integer of man himself, in this infinite sea of energy.

"Man's consciousness therefore, will correspondingly be enlarged with the understanding of the vortical movement of energy of which he is presently both the subject and the object!

"Science therefore, cannot be any greater than the sum of the parts which are presently the ascribed nature of the development of any individual scientist. Hence, science develops with respect to the constant pushing out of the boundary lines of research by those individuals who are involved in the great research of life. Even the scientist, Edison, ascribed the success of his endeavors, to forces that existed outside his conscious mind.

"Nikola Tesla is no doubt one of the greatest examples and proofs of an interdimensional science initiated by his ability to receive information from a dimension that lay beyond the third dimension. He could see the entire minutiae of the electronic devices which he was responsible for bringing into the third dimension. This was proof that the mind functions alternatively in two dimensions. That certain individuals are capable of recognizing this alternate movement of energy is evidence that these individuals contained an awareness of their higher consciousness, that life does not stop at the end of the atomic cycle of the third dimension, or that the physical body does not conclusively close out the consciousness of any individual as a 'fait accompli'!

"Any responsible individual therefore, who would search into the history of science, would discover that such discoveries expanding the boundary line in the quest for the resolution of life, have been earned by those individuals who have recognized the existence of a higher Intelligence. This higher Force was responsible for providing knowledge to that particular inquirer, and establishing a newer boundary line which man was capable of accepting as the extension of the frontiers of knowledge for that time.

"The conclusion of the 20th century is the time for such an advance of the boundary lines of physics and other disciplines, as a result of the work of Max Planck, Albert Einstein, Robert Oppenheimer, Nikola Tesla, and a host of other dedicated souls. We could include among these certain individuals, others in the present who are laboring now to resolve the dichotomies of the basic mathematical calculus and other types of equations to help explain the interdimensional dynamics of energy and the psychodynamics of consciousness.

"Science must be considered to be not only a factor in the area of physics but a factor in the inquiry into the existence of man as a relationship to one another. Science is the extended relationship of man with all factors of nature, because one cannot separate any aspect of life without endangering the species in that separation.

"Evolution can therefore be seen to be a development into the understanding of the physics of the fourth dimension. Everything else pales until that realization is ascertained. The logic and reason of third dimensional life says that man must develop a basis for any statement of life. He must have some basic foundation firm enough to add additional concepts which are attained through incremental associations based on the logic and reason of experiential activity.

"Basic concepts that had been lived with and accepted and that become the security of mankind at any moment in history, must give way to new concepts!

"Albert Einstein has related that he did not accept the conclusive features that had been introduced by quantum mechanics, that indicated that the infinite was of a mechanistic design. This, Einstein completely and irresolutely denied as he stated in his oft quoted remarks, that 'God does not play dice with nature'. In other words, there is no chance, happenstance or accident in the movement of life in which we are both consciously aware and unconsciously involved.

"Energy, if it is a result of an intelligent designer must then, not only in its macrocosmic design, but in its microcosmic design, carry out and reflect the exact principle in which this energy functions. On this basis Einstein validated his awareness of a fourth dimensional proposition. But until other developments could take place such concepts could not be validated in the third dimensional equation in which science was still carrying 'coals to Newcastle'. In this respect, the basic boundary lines of science were still dependent to a great degree on Newtonian physics, even though both Planck and Einstein had proven that the mechanics of energy, as had been determined by Newton's physics, did not operate in the greater infinite!

"Cosmologically, the principles that attempt to explain the expanded nature of the universe and of its causation, are not capable of being understood, using Newtonian physics. The mentality of man had not been completely opened up to the nature of his own great involvement in the pursuance of the principles by which the infinite functions! This is the greater revolution therefore, which is touching upon all aspects of life.

"The greater revolution is 'quantum mechanics'. It has opened up a new field which definitely states that man must himself begin to realize that he contains the basic lever which will open up the door to reveal the alternate movement of energy which is being subjectively seen through electron microscopes and through other research endeavors. Particle accelerators or cyclotrons are describing the inner movement of the energy of the atom. That inner movement will never and cannot ever be explained, using the basic hypothesis of Newtonian physics.

"Quantum mechanics is steering scientists in the right direction. And yet, there are other factors in the pursuance of the truth of the atom, factors from which scientists have turned away.

"On this Earth planet there are relatively few scientists who have had the strength and the belief that there are intelligent forces existing outside of the five physical senses. That source will never be found, unless the individual researcher recognizes that it lies within his own mentality. He cannot see the evidence of his research except through clearer eyes that which is the manifestation of a great electronic force field that encompasses within it, a greater understanding of the life force. This understanding will be of a greater intelligence which will direct that scientific researcher in the proper direction; a direction which will reflect answers to the puzzle in which he is presently involved in his particular research.

"The revolution can then be seen to be a revolution in the extension of man in his field of inquiry, and the realization that he, himself, is the research project that must be mounted. Man must realize that he is the integral field that provides complete knowledge of the extenuating forces that seemingly are inflicted upon the atomic forces of his third dimension. It is a simple proposition, that when one erects a dam he holds the water within the area in which the dam has been built; that water is therefore separated from the area to which it would naturally flow. Consider that this was a basin of water and that a river was the initial carrier of this water. Man must, in this respect, see himself as a factor of the river, of the water in the river, that he is ever receiving the intelligence of the movement of the water molecules and all information that is carried within this carrier wave.

"Individually and as a whole, whoever the individual may be - a housewife, businessman, artisan or scientist, by erecting a dam, mentally speaking, from and around the great movement of energy, which is the resolution of the infinite, he has stopped the flow of the movement of this energy, which carries intelligence from its basic initial thrust. It is that energy which he lacks which is the block in his aborted attempts to gain greater understanding of the enigma of the atom.

"Hence, whoever may be the inquirer after truth and whatever the proposition, the inquirer must also accept the fact that he blocks himself from the truth or the answers to his basic questions which he is proposing. The answer can never be given to him satisfactorily because it will never gain an entrance into his mind if he does not recognize that he is responsible for the dam that he has erected as a result of previous experiential developments of his life. The basic truth, therefore, is that man has lived before!

"Scientists who are thus pursuing a course of investigation into the anomalies of the universe, and into theories that have to do with the creation of the universe, must realize that they cannot further the basic propositions of their research unless they add to this proposition that they have lived before and have accumulated a vast debris of information which has presently become the dam! This has closed down the avenues of information which could become a development and which would give them a clearer rendition of the basic anomaly and enigma of life.

"The pieces will fall into place, when man realizes his present position,- that he cannot receive a consistent picture of the history of man or a consistent picture of the movement of energy in all of its factors that affect planets and man, if he does not realize the continuity principle of life! Moreover, not only is the mind a property of, and a regular continuity of the fourth dimension, but it is a feature of the entire evolutionary design of life, with man the integral product. This evolutionary design, an intelligence which is one's electromagnetic instrument, is the mind!

"Of course, there are those who strain at the bit and cannot necessarily extend their consciousness to the point where they can conceive of the immeasurable and incalculable nature of infinite intelligence. This is the design of life of man, in his own measure, through his own experiential life, from life to life.

"Experientially then, no man is similar to another. Individuals have developed according to their own ability to conceive and to develop the necessary instrumentation, which is the logistics of their mind, to be able to move to the further shore. Therefore, there is, in this awareness, the recognition that within the community of scientists there are those who are further ahead in their ability to conceive of the further boundary that separates the third dimension from the fourth dimension and where the laws of third dimensional physics are no longer operative.

"This is a basic factor of the change in the 20th century, from Newtonian physics to fourth dimensional physics. It resolves itself within the operative movement of energy as a force moving on a force contained within the boundary line of the speed of light, but is no longer capable of explaining the energy that is seen to function in the subatomic particles. It is this recognition that the boundary line of the third dimension is being reached and that the fourth dimension is being seen. As the ship's captain receives information from the watcher on the mast, who sights land ahead, so are those scientists who are truly the pioneers in the extension of knowledge of life in the atomic world of the third dimension, calling out that they 'see a distant shore'. But they could not possibly identify the distant shore unless they themselves, conceived the physics of this fourth dimension which is the farther shore and which is man's destiny into the evolutionary design of infinite intelligence.

"Man must realize that he is a developmental factor of the infinite, so ordained to constantly expand in the selfsame manner in which he, as an energy system, a vessel on the greater sea of life, must navigate his instrument so that he may reach a farther shore. Then upon reaching this farther shore and putting something upon it, and developing a new way of life, he will, in the future, so to speak, find himself advancing into another boundary or shore that will recede as he realizes that he must constantly add to the nature of his understanding of life, and that he must constantly evolve. This is progressive evolution! It is progressive evolution that is presently making its earmarks within the life of man.

"Since science is the basis for a progressive evolutionary development, science is always itself mirroring the advances that become a realizable factor in a progressive civilization that will be lived by those peoples who have been capable of harmonically attuning to these changes.

"After all, all changes of life are not mechanistic. They are the interrelated factors that have been developed by mankind himself. The computer revolution is a development of this realization of a higher physics. This higher physics is a recognition of a power that functions free from the barrier of time and space.

"The new physics must be the means through which the new frontier will be pioneered, as man has always desired to constantly extend his present environment, mentally and physically. It is the mental environment which first must be updated before his new physical environment will, as a result, be the resolution of a new age. The new physics therefore is this: It is the removal of the dam whose creation has been a slow process of accumulation and which has also been slowly broken through. The hammering upon the rigid walls of this dam, have been the efforts of those strong individuals, Kepler, Newton, Nikola Tesla, Planck, Bohr, and of course, Einstein and other individuals who are beginning to realize that they can no longer depend upon the mathematics that had previously provided them with the premises for the picture of the classical movement of energy.

"The nature of the atom and the nature of the universe and the nature of man is a balance between the elements. No understanding of the microcosm or of the macrocosm can be developed without the understanding of the integral factor which is man, who is so balanced between the two forces to contain the complete unified nature of the both. Man is both the microcosm and the macrocosm. It is this study that is the basis of the new science.

"The new science of life which integrates all of the behavioral methods of scientific research, integrates all fields of knowledge and disciplines, has been the pursuance of knowledge as to the understanding of man as an integer of the infinite. This science will reveal that man is a product of this infinite and contains within his integer the whole number, which is the unified field force that is his higher mind. The superconsciousness is this unified field. But as all unified fields contain also participles of itself to validate its existence, there is the subconscious which, to all extents, is a unified field, but a subsidiary of the superconscious self.

"Fermat's theory can therefore be understood only

within relationships that are the continuity of life of man in his physical anatomy - from his fourth dimensional field to his third dimensional field. Therefore, the subconscious which provides the recognition of his present ability to exist within the limitation of the light barrier and within the field equation of the sixteen to twenty elements that make up his physical anatomy is, to all intents and purposes the causality of life within that frame of reference.

"But man is also a whole note. It is this wholeness which is his infinite creative nature that is fully realizable and functionable as his superconscious self; the sum and total of all polarized experiences which he has accumulated in his previous lifetimes through the subconscious.

"The energy of man is divisible so that he is existing within the integer that provides him with the security of an atomic environment. But he also exists within another security which is non-atomic and is the whole number, the true whole number because it is not divisible. It is a continuity in a positive polarity relationship to all factors of life which are the extension of it in its evolutionary progression into the heart of the infinite itself, if we can use that basic metaphor.

"The infinite can only be understood through the development of the transposition of itself into the integers which are man's consciousness, gained through development in the atomic field of life. No one can possibly explain infinity when he is still in an elementary grade capable of conceiving values, yet finite and which do not admit in their conception, the existence of its infinite nature! Man is therefore a developmental relationship of the expression of the infinite, and this Infinite can only be understood when it exists within the proposition of man's evolutionary development.

"One is therefore constantly researching the evidence of infinite intelligence within the periphery and the boundary line that is man's present consciousness. And as this present consciousness is only one of the tributaries of the infinite, and as the tributaries are so navigated by man, it will open up into another tributary until, through the exercise of logic and reason, man is thus joined in a greater fashion to the body of water to which this river is flowing. That is infinite creative intelligence.

"The fourth dimension is not a singular arithmetic factor to be worked with, using the symbology of calculus or geometry or any of the third dimensional equations. It is an equation that can only properly be applied by the extension of the present boundaries which are the foundation for the stepping stone of man as he is moving into the higher field - a field in which his mind will so advance and expand that he will, as he raises his binoculars, see into a vast distance and realize that there is no distance, time or space, in the infinite. All is frequency! Time is a postulation that has been a development for the elementary expression of the products of the infinite which is Homo sapiens.

"The evolutionary design of life is therefore a fourth dimensional proposition, which further advances the reality of life into higher equations. But man is now reaching that point where the perspective of a fourth dimensional reality will become his understanding of life, and where man's consciousness will have lessened its dependency on the time and space dimension. This is man's 21st century development, the new age of reason, the new age of logic, the *new age of spiritual renaissance.* This is because man is a *spiritual being.*

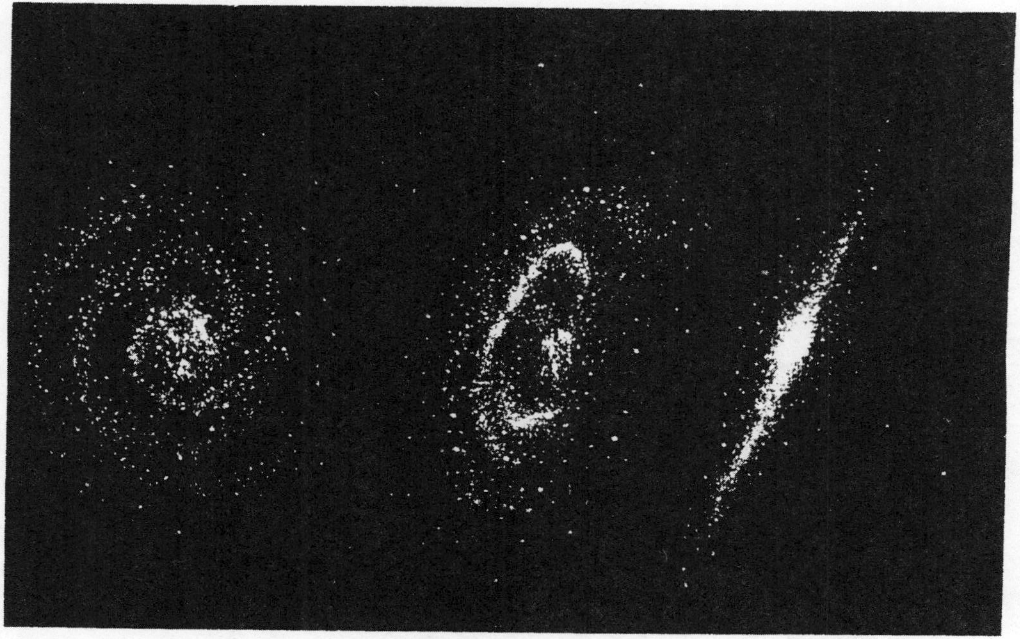

Portrait of galaxies in the vastness of cosmic space.

Chapter 3

AN INFINITE CONCEPT OF COSMIC REALITY

"Let us say that life is a constant proving and qualifying relationship. This relationship has always been within the quantum which is man's own mind. The basic rudimentary physics of life with any individual is his desire to understand the nature of the reality he has accepted as the secure boundary line of life. The mind is a quantum and an integral factor of the continuity of energy, and is, in its own stratum a functionable intelligence of a creative design. Yet, many individuals have remained within a boundary line in which they have accepted certain limitations as a way of life through language, time, and space.

"Let us therefore examine the true meaning of time and space and we will see that one is securely interwoven with the other. There cannot be time without space, and space cannot exist without time. The meaning of this geometry is no credit to those who have used the symbolism of third dimensional mathematics. Mathematics is not necessarily incorrect; that is to say, mathematics is the reflection of an individual's mental understanding of the energy within which he exists, of the macrocosm and the microcosm. It is the form and function of mathematics that we are alluding to that must be developed if there is the desire - and there should be - for a progressive evolution; an understanding of new boundary lines which make necessary a new mathematics for the understanding of infinity, a fourth dimensional paradigm!

"The reality of life's existence within the plasma of a third dimensional framework, within the boundary of the speed of light and within the known sixteen elements of the physical anatomy, and upon the supposed firm foundation of a planet's surface, is evidence of energy. However the description of this energy has gained its credibility, only through the existence of five electronic devices called the five senses. These five electronic devices are a recognition of the existence of the physical brain from which these senses are derived.

"Therefore, all evidence of the motion of energy as it exists within the framework of man's five senses has been so derived as a certain geometry throughout evolutionary periods from the known histories of man, the astronomy of the Egyptians and of the Greeks, and up unto the time of Kepler and continuing with Leibnitz! He, along with Newton, is responsible for the introduction of the mathematics of calculus as a symbolic technique for deriving the answer to some philosophical questions which have become a quality of deductive reasoning. Such reference points are helpful in the application of the logarithms of sines and cosines which are the basic factors in the application of these algebraic symbols.

"However, these basic symbols are only relative to the movement of the energy that is the derivative of atomic life within the time and space dimension. Moving out of the space/time dimension into the micro energy fields of subatomic forces, we find that time and space are no longer relevant. Hence, it is not correct to extrapolate the nature of energy in such framework that is no longer coexistent with reference points within which man is regulated, in his measurement of time and space.

"The very cell structures of man's atomic body are geared to its relationship in the space/time dimension in which it functions. Time and space are therefore the regular features of man's directional ability to survive and to develop the understanding of himself as an ego structure, a necessary development in the cellular nature of each organism in the grand sweep of evolution. Yet man is more than an organism.

"Scientists are individuals who have attained to a particular development in their desire to fashion a clearer picture of life by applying mental concepts that are the highest derivative of man's evolutionary development. These mental concepts come down to a basic equation, logic and reason. However, logic and reason can function quite well within the parameters of a third dimension and quite well within a society that has already regulated itself under the jurisdiction of certain laws and customs of life. The business community has a parameter in which the businessman functions under a regulatory system. The farmer functions under a regulatory system which provides him with the basis of sowing his crops and harvesting them. The teacher in the class room functions within a regulation which is the nature of his society and the needs of his students and his own evident interest in opening up the minds of his charges.

"In every way we see there is regulation. But the regulation is one of 'man-made' properties which have been necessarily derived from the field, the energy field, of course, in which man functions. This energy field, based upon the quotients of a space/time factor, has a derived beginning and an ending. Up to the present time in man's evolutionary beginnings, all questions, even of a cosmological nature, are still related in this space/time, beginning and ending schematic.

"Hence, even the determination of the creation of the

universe, is based on the same proposition of a beginning and an ending; the postulation or theory that life began with a 'big bang', a huge atom that exploded with great force and moved out into the space about it. Presently, the seen and unseen galaxies, and other anomalies are a puzzle to astronomers searching for the appearance of strange energy forms that populate the universe.

"We return to the proposition that man contains the key to unlock the mysteries of life. Time and space are factors of a dimension. The understanding of life cannot be furthered by a singular awareness in ones present environment, because it requires man's continuous desire to see through the third dimensional boundary line. When one begins to conjecture on the system of time and space, within which life is regulated through clocks and the biorhythm of cell structures, and the regulation of societal life, man will recognize his interdimensional nature. The exterior factors of the physical dimension cannot provide the answers to the interdimensional movement of life's energy.

"The interdimensional movement of life cannot be contained within the systems of present calculus and geometry! A new mathematics must be discovered! Certain scientists have mentally escaped time and space, in the recognition that energy no longer behaves according to classical physics; that it no longer obeys the regularity of light traveling at 186,000 miles per second. It shall be discovered that light does truly oscillate at 210,000 miles per second because of the imperfect nature of the measurement of the electromagnetic plasma. Light does not travel in a straight line, and hence it cannot be correctly measured through a length of pipe one mile long.

"Light is considered the boundary line beyond which

there is no life. Light is the very channel through which man gestates the developmental factor of his intelligence. This gestation period is being completed at the end of the 20th century and the completion of the sixth cycle of the recessional. This sixth cycle carries within it, the sum and total of all past life experiences which live within the cell structures of each individual, physically and mentally. The seventh cycle is discharging into mankind, Homo sapiens, the nature of a higher expressionary intelligence.

"This is a higher frequency band of energy that becomes the nature of all things that live within its higher cyclic nature. As all energy is coexistent with other energy structures, there is in this respect, the movement of man as a nuclear form of energy, moving up from the compacted third dimensional nucleus of the vortex, into a higher energy band within the interdimensional vortex. Man has moved through the barrier that once was considered to be the end of life. He has moved beyond the bend of the river, moved beyond the speed of light, and is no longer held within the framework of an atomic form!

"Naturally this is an understanding that must be prefaced by all other factors in the development in any one individual with respect to his own mental discipline and the ability for him to realize the inherent nature of his own mind. When one has been transcended in certain moments of life's experiences where time, to all intents and purposes has been compacted or shortened and space has dissolved, then man becomes a bird and has found a freedom of excessive beauty. This experience can no longer be considered physical because it exists on another energy level known as psychic or mental.

"This mental experience is not only non-physical but is

removed from the boundaries which are the regulatory means of living within the physical anatomy. It is truly a realization of the existence of the continuity of life on a higher frequency band where time and space no longer exists. The reason for this phenomenon can be substantiated in that man is not a physical creature.

"Man, by the very hieroglyphics of that three letter word is an energy system; an energy system functioning albeit within the measurement of a third dimensional field of energy.

"He thus is an energy system, always functioning within a larger energy system and therefore needs to have an instrument in which to contain this functioning design. This energy system, as it is infinite, is timeless and spaceless, and must have some, shall we call it, restrictive device or instrument which will allow man to navigate in a lower frequency field. In this respect the physical anatomy is a derivative of the atomic material of the third dimension. Hence, man has always been puzzled by the dichotomy of himself. On the one hand, he is affected by a five-sense body; on the other hand, he is also affected by his sixth, seventh, and eighth sense body etc!

"Putting it in another way, man is affected by his physical anatomy and his spiritual anatomy. His physical anatomy is bounded by time and space; his spiritual anatomy is a higher energy system that correspondingly functions on the fourth dimension. So in this respect, it can be substantiated that man is a fourth dimensional energy system, functioning in a third dimensional restrictive environment within a five-sensing electronic instrument.

"The entire dichotomy of life can be resolved then in the understanding that man is functioning through an instrument that can only express with the elementary five senses, to learn in this elementary fashion of certain factors of life. These factors are the nature of the physics of the third dimension. Yet he must learn in this manner the physics of his fourth dimensional mind. The boundary of atomic life within the third dimension is a development to a greater awareness of the boundless and infinite nature of energy in which all mankind has been created as a spark of infinite intelligence.

"The Infinite Creative Intelligence, the Cosmic Mind, the Fountainhead, known by names that have been garnered from various teachers and teachings, the Father, the Infinite Force, and so on and so forth - names which have, in some respects, advanced the understanding of man and in other ways, restricted his infinite nature. The carrier wave that contains the logic and reason of the interdimensional nature of man exists within the logical methods of scientific development. But that development is not new, because the existence of this knowledge had already been a developed feature within the existence of man on other planetary systems within the universe.

"Why this has not been a recognized and accepted factor in the present 20th century is man's own determination and responsibility to realize, for he carries the impacted nature of his own mental blocks, a surrealism which has clouded the truth of his own reality!

"The continuity of energy, by its own acceptance, is an indication that life has existed previously; not only has man existed previously, but that he is a continuity that contains a reversible and progressive nature. The interior design of energy is an indication of the movement of man and a movement of all factors of life that are related to the existence of atomic organisms from positive to negative polarities.

"Science and scientists therefore are the validating means for helping struggling mankind to realize the potential of their own minds. They can further the development of mankind, progressively, by pointing them in the direction of fourth dimensional physics. And you may ask, how can fourth dimensional physics function in a third dimensional world? We answer in this fashion: Space/time is the functioning environment through which man exists in his physical anatomy and yet man is a fourth dimensional factor of the infinite. He has always been directing his fourth dimensional self into the activities of his physical dimension. Unbeknown to himself, he has been constructing subliminally the reality of the interdimensional principles through which intelligence develops to a greater potential the logic and reason of life itself!

"The greater recognition of the higher dimensional nature of man will give a strength to individuals who have not yet arrived at the point where they are capable of logically deducing the interior nature of their mind. Particle physicists are just now deducing the interior nature of the greater intelligence that is pressing down frequency-wise into the nucleus of the atom.

"Scientists can therefore, advance the reality of life as a reflection from the higher dimensions; and the validation that life is a result of a higher Intelligence contained in the causal dimensions. The birthing of man is a design of infinite intelligence, to give man the necessary so-called time, which is a cyclic movement to develop experiences. Such experiences may be in a span of so-called earth time, millions of years, but are fourth dimensionally, a matter of moments. How long it takes for an individual to realize the Reality of himself as a participating element within the greater whole is of no consequence. It is the realization and the recognition that is of importance.

"Therefore, the fourth dimension is the higher mathematics, to be understood and developed so that it may function as navigational lights, that have been lit by individuals who have been previously trained in these matters. They will therefore function as guides, to help those who are young souls, who are coming along in their evolutionary development.

"And so science is, in every respect, the navigational beam which, when properly attuned to the transmitter, will be able to reflect the knowledge of the interdimensional nature of man and of life. The navigational lights will therefore give mankind the ability to move into and across the barrier of light. Man will then be developing the realization of the existence of certain fourth dimensional factors, equations he is learning while living in the physical anatomy. When he arrives on the other side of the third dimension of atomic matter, the energy of momentum, inertia and gravity will no longer hold him captive within its force field. Thus the individual who has arrived at that point in his evolution will function beyond the mass spectrum of the atomic universe where light oscillates on a higher frequency. Man will be able to function within new boundary lines which will be a reformation in his present consciousness.

"Light is now moving beyond the velocity of 186,200 miles per second. The wholeness that he will experience will be without references to a beginning or an ending. He will always know that he is a consistent whole and will not be elementarily concerned with a start or a finish. The perception and conceptualizations involved within the individual in such an ability are the developmental factors that have been attained through life-to-life experiences in the space-time continuum of atomic and non-atomic worlds.

"Space and time are the nature of the schoolroom, the educational facility in which the rudimentary knowledge of evolutionary life is being taught, for which all factors of life lived through the physical anatomy are basic teachings. Earth life can therefore be seen to be the basic curriculum for the graduation of the species man, into the next and higher classroom, a higher dimensional classroom which contains advanced concepts as to the meaning of life's expansion!

"As the student of life is directed from his childhood, from kindergarten to primary school, to secondary school, and from secondary school to colleges and his university classes, so does he attain to the postgraduate classes where the basic rudimentary knowledge of a particular aspect of life is expanded. In this expansion it takes on more complex factors which exist within itself. In the same sense, man will advance into the postgraduate classes where he will be able to deal with the fourth dimensional principles that exist within his own mind.

"This mind is an anatomy that is constructed of sinusoidal wave forms, functioning on a timeless, spaceless structure, which has been extrapolated from the atomic dimensional senses, yet functioning in a cyclic manner, unbounded by time and space. This is the present existence of man in his postgraduate development.

"To qualify and successfully complete these examinations and to occupy a higher station in life, one must determine one's ability to answer the posed questions. In this respect, one eventually loses concern for self, to begin to become a spiritual being who has graduated from the elementary particle field that is the atomic feature of life!

"Should you not be capable of advancing to the post-graduate classes in life's learning, you will necessarily then, have to take more classes. That is to say, you will necessarily have to live and experience lives lived through the physical anatomy, and in this respect, you will be advancing the nature of your spiritual reality but on a lower spectrum of the infinite.

"It is the present nature of life, derived through man's previous existences that have prepared those who are now qualifying for these examinations. These individuals will become the teachers and guides for the New Age of Logic and Reason, the Renaissance of the mind; the true rebirthing of fourth dimensional factors of life.

"So long as the reality of time and space is an accepted feature and mode of life, the individual, although he contains the fourth dimensional equivalence of energy structures, will be funneling this energy into his restricted atomic nature. He will therefore change, or shall we say, he will sublimate the expanded nature of the existence of energy as the proposition for his progressive ability to disengage himself from the gravitational pull of an earth world.

"Man can live in his physical world in his society while developing the expanded knowledge of his fourth dimensional nature, because on the very basis that he is stepping up to a higher intelligence in his realization of the unbounded nature of himself, he can look down upon the lower threshold without being tied to it. In other words, he can be *in the world but not of it,* and in this respect, he can help his brothers. He can help mankind by advancing the higher or more intelligent factors of the interdimensional nature of life to those who are still students and have not yet gained sufficient experience to develop the mathematics of their own minds in the logic and the reason that will give them the necessary ability to determine the reality of this abstract nature of themselves.

"Those who are in the world but not of it have abstracted from the infinite a sufficiency of knowledge so conceived by them that it becomes the basis of their own mathematics; a basis in which they can breathe the higher intelligence of the fourth dimension and breathe the atomic substance of a third dimension. To this extent, since it is a delicate proposition of knowledge and usage, it will provide them with the balancing ability to exist in a physical anatomy but to live with a non-physical mentality.

"Of course one cannot live in a higher dimension while in the third dimension except in certain moments or certain times, shall we say, or cyclic relationships for so-called short durations. But the fact is that man is learning of the existence of the fourth dimension by having the experienced transcended states of consciousness. It is transcendent because physical laws which are the boundary lines of the gravitational field of the third dimension have been transcended. One must learn therefore of the mind as an energy system or else how would one be capable of existing in an environment where time and space is nonexistent, where the individual lives in a non-atomic body? The ability, the possibility, is the potential held out to homo sapiens!

"So the third dimension is a classroom for man to learn to regulate his life from two alternate perspectives. First, man must learn to live and master his physical anatomy. He must learn to master the evidence of the interaction of himself in the physical body in the society of other peoples who are likewise functioning in this manner. Therefore, in learning and mastering the third dimensional factors of positiveness and persistence, he will attain a greater intelligence so that he can eventually function from his mental self!

"We now come to the true recognition that life is a development and a constant mastering of various factors of life which are the complexities of that environment. In the mastery of the physical evidence of life, man is capable of entering into a new environment where life takes on a new proposition, to travel on a boat that has no sails, no wind, but yet must be advanced within the sea of energy. This is his new body.

"Hence, the individual learns that he is truly a nonphysical individual or being, that the energy which he had thought to be of a physical nature was, at all times, a fourth dimensional factor and that at all times he had been learning about the higher principles of the energy of life. Thus life is a preparation to provide each and every individual the ability not only to exist in a higher framework of life but to advance the entire proposition of life. As each individual advances in his mastery of his life, he so advances life of other factors of his society.

"Society as a whole then becomes an integrated part, and an expanded nature of the infinite. The movement forward in new and higher equivalents of the understanding of the inherent complexity of Infinite Intelligence is the expanded factor in which this Infinite has, at all times, taken on its form. Therefore, as man is the microcosm of the greater macrocosm of infinite intelligence, by his forward direction, in the expansion of his awareness of life and of his intelligence, he is becoming the macrocosm, taking on the greater intelligence that is the inherent nature of the energy structures, the expanded lines of force that make up the universe. Space/time is thus the rudimentary factors and the developmental beginnings of mankind as an infinite relationship to his own progressive development.

Chapter 4

THE ENIGMA OF THE PLASMA UNIVERSE

"We have reached an age in time when the real nature of the mind must come into a higher recognition, and put into the 'limbo of the past' the many conflicting theories that attempted to provide some understanding for the reason of life, and the place of the individual in the universe. Naturally, these kinds of questions cannot be answered in one book; yet, the answers to those questions are contained in the library of man's past life experiences.

"Long ago before the tragedy of Atlantis, there was a concise bibliography, and in fact, an esoteric foundation of information that traced man's heritage back throughout the ages of Atlantean civilization, and beyond that time into Lemuria, and even beyond that, when man had first lived on other planets. This prehistory of man did contain recorded writings by individuals, who, in fact, had a 'real' understanding of the interdimensional nature of man, and of the atom as they exist oscillating between the third and higher dimensions. This is the complimentary cohabitation of physical man with spiritual man!

"All interdimensional life forms must live in some cohabitation! There must be members of different interdimensional societies who mingle with each other, so that this intelligence can be transmitted to individuals who are just beginning to understand the *continuity of their consciousness.* They are learning of the continuity of themselves as real participants in a *life-system* that is circular in its structure. It is not only unlimited, but it is infinite. Yet, in this infinite system, this infinite cosmos, there must be some beginning in spite of the fact that there is no beginning.

"There must be some beginning in terms of the individual's ability to determine intelligence within himself. That beginning of course, began back in the middle of so-called time. For the infant, it is a beginning. Yet the riddle of the infant is that he is an infant only physically speaking, because he has lived in a mature manner some place, somewhere in some form. But that manner and that form is a new beginning. So, the riddle of beginnings and endings must be placed on its proper level, within the receptors of the individual's understanding of the inconsistencies of one dimension when it is not coupled to another, and particularly, when that dimension is a third dimension.

"A clearer understanding of the cosmic plasma universe, will help resolve the dilemma within each individual, of the fear of death, the fear of fear, etc., and erect in its place, a firm foundation based upon the reality of the plasma field and its electromagnetic properties, which in fact are properties of the unified field force constructed from higher dimensions of the infinite..

"However, each individual must develop the plasma of his own electronic nature as a real factor of the interdimensional nature of intelligence! These are the formative steps for each individual taken as he validates and proves to himself his ability to develop his picture of life on the electronic sensing plate of his mind.

"But let us speak on the cosmological principles of the plasma universe of this third dimension. The following information has not yet been factualized within the domains of the scientific community. Yet certain facets of this information have touched the peripheral consciousness of noted and unknown scientists in the disciplines of astrophysics, cosmology, particle physics, plasma physics, and in other disciplines of biophysics, biochemistry and neurobiology. These scientists have demonstrated through their own evolutionary development that they have not been satisfied with conventional sterile thinking during their physical tenure on earth.

"Theories must be demonstrated to be believable, and must be repeatable in the laboratory. Such theories must follow the logical means of scientific investigation because it is in this manner, and in this way that present principles of electronics and physics have gained acceptance and have enriched the technological side of society! However, the technology has not enriched the human side of society nor advanced his mental constructs of logic and reason.! This is the enigma, and this is the danger!

"We did indicate that technology has provided certain information, and as building blocks, erected a foundation which are the great cities on the planet Earth. However, other technologies have attained a rampant nature in which consumer products and the products of war are competing each with the other, and have little relationship to the spirit of billions of people who live on the planet's floor. Although they may wear the outward accouterments of technology and exercise their mobility in aircraft, and automobile etc., they have not yet exercised the elasticity of the mind which is the motivating plasma of the fixed system of the biological anatomy.

"There is then, an enigma in the life of man, in that he nonchalantly takes for granted his present life in society, and does not understand the dynamic processes of the mind and the brain structures to advance his spiritual nature. He sees to it that his physical body is cleaned and nourished, but does not know of the enervating source for his biological system! He does not know of the cellular nature of his psychic anatomy, nor does he understand the alternate statement of its cosmological nature.

"The past hundred years have placed in the encyclopedias and in the libraries of man, dissertations by men and women who have braved the tenets of society, and have introduced advanced concepts of life. These individuals have lived in advanced social systems where man had attained a coherent understanding of the real nature of themselves as a unified energy system in tune with the infinite. Thousands of years of man's writings contain a complete and total statement by individuals who have spoken of the reality of life, and yet, their discourses have not been collated in the language of present-day 20th Century science. These master scientists have been the forefathers and as coefficients of life, have rubbed shoulders with their younger brothers in their attempt to turn their eyes upward and inward, so that they would see the alternate nature of themselves. Their names are in the halls of fame, they are the *great philosophers of old*!

"Particularly, we would speak of the latter one hundred or so years and of the persons Benedict Spinoza and Rene Descartes, because these two advanced minds spoke from alternate perspectives. Spinoza spoke as a cosmologist explaining life in the atomic plane as a relationship to the interpenetrating life lived by the same individual on its alternate polarity in a higher dimension. Rene Descartes spoke of the inner nature of man, his mental intelligence living through the bioplasma of an atomic body!

"Let us take another side road and touch upon certain facets of an interdimensional physics that once understood, that is to say, perceived within the individual's understanding, will lay to rest the factualism that is the present history of modern science. We speak particularly not only of physics, but of biology and also the 'digger's up' of old civilizations that have now been closed from the view of man, that lie deep below the earth. The impossible therefore, is possible, when the outer limits are recognized as being a limiting factor in the understanding of that researcher into whatever field he is presently concerned.

"In the proposition of the 'big bang' as the beginning for the creation of all atomic forms in the present known universe, there are anomalies that are only beginning to be recognized. By what principle can we explain the creation of galaxies and the myriad of suns and planets that live within their embrace. And what of the dark matter of the universe? What of the immense luminosities as flares that are emitted from the center of galaxies? And what of the emission of light greater than a thousand suns that are thought to be quasars? And what is the explanation for the emission of strong neutrinos and gamma rays from super nova explosions?

"Up to the present time in 1989, the known boundary line of the universe has been pushed back and that which could not be seen has been photographed by ultra-sensitive cameras through optical telescopes. It appears from these pictures that there is constant activity. The outer limits so-called, of the known universe reveal great violent movements of plasma, indicating the birthing of galaxies. The Orion nebula reveals the birthing of new stars. This nebulosity is in fact a plasma, an electromagnetic force field forming circuits in a vortical pattern. The present theory advanced by the Swedish astrophysicist, Hannes Alfven, is the closest yet to the electromagnetic principle of cosmic creation!

"The Big Bang *hot house*, if you will, did not erupt by itself out of a vacuum. Of course, no one believes that there is a vacuum, because there must be that logical *principle of cause and effect;* there must be something to form something else! A Belgium priest first made the statement not more than fifty years ago, that all known and unknown life forms were created from one singular atom which contained the synthesis of all that was necessary for the development of the universe and the population that evolved with it! However, the theory has many faults; it has not explained the factual cause for the clustering of galaxies, and the tremendous areas of dark matter in-between!

"In an explosion, it is well known that the energy emitted is coherent and spatially connected, so that there would be an equal movement of energy from all parts of the originating explosion. For this reason star clusters do not fit into this theory. Dark matter and the elliptical orbits of galaxies do not fit into this theory. The expansionary principle of the universe, the Hubble model is based upon the red shift and Doppler effect. And here, we take you back to our first statement of the primitive measurements which man uses to gain an understanding of what he sees, billions of light years from his Earth planet, which have not been capable of reflecting the *real image of the universe!* And the fault lies in the spectrograph!

"A great and mistaken fallacy is the red shift. On a spectrographic analysis, scientists compare the appearance of certain lines as they appear on the surface of the refractory, crystal prism. Like an ordinary triangular crystal of glass, the spectrograph divides the light spectrum into certain lines somewhat similar to a rainbow. It is assumed that all known elements found on earth are part of every star sun being viewed, as these are similar to our own sun. The thermonuclear process, or burning (fusion) of these elements, radiates the necessary light viewed on the spectroscope. From very distant star suns, the red light appears at a different place than it does here on earth. The scientist thinks that this is caused by what he calls the Doppler effect; i.e., the train whistle receding away from you has a lower pitch than it does coming towards you. The scientist thinks the star sun is moving rapidly away from the solar system and the galaxy, because he is always looking out towards the outer rim of the galaxy, and he cannot see towards the center because of the many more star suns and other conditions. He must always look out into the Milky Way; so therefore, he thinks that our galaxy and universe is expanding. This leads to the most ridiculous of all assumptions - the big bang theory; i.e., somewhere at some time, a huge ball of atoms collected together, and when there were enough of them, and they were compressed tightly enough, they suddenly exploded! The little pieces traveling outward formed the countless trillions of star suns of the universe.

"Now, scientists have not yet explained why these little pieces, in traveling out, assumed that pinwheel fashion, or the rotating formation with which we are familiar from astronomical photographs. Logically, from an explosion, such objects would be propelled straight out from the center. Also not explained: why the universe is thicker at the center and tapers out to a thin edge. An explosion should hurl bits and pieces in all directions, up and down as well as sideways (360 degrees). Also not explained: why countless galaxies have formed in this universe, each one rotating pinwheel-like, which again invalidates the big bang theory for the creation of third dimensional universes.

"Pictures of our sister galaxy in Andromeda prove the existence of these galaxies - impossible to be formed in an explosion. What has formed the universe and these galaxies has been adequately described in the book, *The Infinite Concept of Cosmic Creation.* A huge interdimensional centrifuge formed this universe - the galaxies being primary anomalies or eddy currents, generated from the net sum and total of this electromagnetic energy in this centrifuge. Here again, this same harmonic regeneration takes place in forming suns, planets, planetary systems and atom-energy forms, again, all remanifesting in their respective weights, the different electromagnetic energy and cyclic wave form patterns from their own particular vortexes or centrifuges. The net sum and total of all these interdimensionally regenerating, electromagnetic energy force fields prove how wrong the red shift, Doppler theory is.

"Einstein's mathematics showed how light is bent around intense magnetic fields. Energy, as light coming from a distant star, is curved by the electromagnetic lines of force of the galaxy. In accordance with the respective frequencies involved in the difference of light frequencies, the effect is to bend the shorter, or blue rays, more than the red rays; the yellow rays falling in-between.

"So, the scientist makes this big mistake: In his spectroscope he places his comparative graph lines to line up first with the blue light ray lines, and the farther out on the scale he goes, the farther these lines appear to be out of alignment with his comparative chart, which was made on earth, lines which were photographed and made into a picturegraph at a very short distance from their source, and were bent by leaving electromagnetic fields. So, the red rays would naturally appear at a different place, and farther out on the spectroscopic graph picture - not because that distant star sun was traveling at incredible speeds away from the center of the universe, but because just like with a prism, the light rays were bent to form an arc, and following the curved radial lines of the galaxy, the shorter blue rays were bent slightly more than were the red rays. The same process was duplicated by the spectroscope, but as yet, this stupid blunder has gone unnoticed.

"Only one more fact should be noted: that all energy rays which are coming from our sun or from any distant sun are really not heat or light, until they are converted in some relative degree by the electromagnetic fields of the earth. Again, this conversion takes place according to the total frequencies involved; heat coming out as one subharmonic, while our three primary light frequencies are higher harmonic frequencies.

"Einstein's theory of Relativity, has shown that the universe is curved! That should have been the first clue to astronomers, astrophysicists and cosmologists. That should have been the first clue, as light is bent as it moves from star to star, and galaxy to galaxy! The universe is a spheroid! Therefore, an astronomer as he peers through his telescope or radio telescope, is seeing objects at a distance, but he has no way of determining whether these astronomical objects are moving away from his initial observation. His only determination for factual evidence, is in his use of the spectrograph where he can determine within the locality of his own solar system, the distance between planet earth, the sun and the eight planets that surround the sun. However, in the determination of the distance between other galaxies, the error is applying the measuring device of a spectrograph.

"Since the universe is curved, the spectral lines of force that have been drawn on the spectrograph do not coincide, and in fact give an appearance that is completely opposite, factually, to the true reality. Putting it to you in this manner, then the galaxies are not receding. The universe is not finite, but is infinite. The universe was its present size and ever will be its present size. The universe is more than what is capable of being determined by present instrumentation. Until present instrumentation is changed to reflect the real property of the electromagnetic field of the universe, there will not be the ability to cross over the barrier that has limited the man of physics, who is not yet willing to accept the evolutionary design of life.

"But let us talk about the universe, and realize that the present universe is an imprint, an image that is constantly regenerating from the parent from which it was birthed. The universe, yes, is developing; it is constantly going through a series of paroxysms. It is in some respects, similar to changes that take place on earth planets, and on suns, where geological processes are constantly taking place within the center of planets and suns, and galaxies. The universe contains the emerging life forms that will provide some basis for some evolutionary development of mineral, plant, animals and man. The universe is a tremendous giant plasma body; an infinite design of an electromagnetic nature. This electromagnetic nature contains the imprint of higher couplings from a dimension that is the superior or the primary motivating cause. Then we see that the movement within plasma has a purpose.

"The vortical nature of the plasma field as it oscillates alternately between its magnetic and electronic nature, develops a resonance, so that the oscillating plasma finally resolves into the known galactic spiral arms. These spiral arms are the genesis, so to speak, of suns and their planetary bodies. The revolution of these vortices within the galactic arms, are resonating from the motor, so to speak, that has generated this effect from a fourth dimensional plane, and in fact, is a facsimile of the movement of this fourth dimensional plane. The universe is a stable body, constantly regenerating new suns within new galaxies. These astral bodies are regenerating through-out, so-called space, as the new basis for conscious life.

"A sun that becomes destabilized based on the anachronistic changes within the electromagnetic forces explodes, short circuiting its energy system over millions and billions of years. There is nothing lost in the universe. That which seems to be a remnant of a dying sun is the beginning of a new sun and is exchanging its rich nature of its component energy where it is needed.

"That a singular atom exploded from nowhere to provide the basis for the life of Homo sapiens is full of inconsistencies. It is inconsistent, primarily because it is not logical. Nor does it provide for man's ability to conduct his life so that he would enrich society and the evolutionary mandate of life.

"We must speak the truth of the reality of life as an interdimensional movement of electrical energies. This is in fact, the reality of the real body of man, the real body of the neuro-synaptic nature of the brain to activate the cellular, the atomic body in his motor and vascular system. Therefore, to rise above these inconsistencies, we must sever any limitation that bends the mentality of mankind in the reality of life.

"The cellular body itself, must have a relationship to its non-cellular body. That is to say, the physical anatomy must have some relationship to its non-atomic energy system. The physical anatomy is an outgrowth of the chemical system of the universe. In some respects they are really one and the same. We can see that man is an evolutionary development of chemical elements derived from the electromagnetic energy spectrum. In fact, as it has been proven in laboratories, the chemical elements of the sun and of other planets are identical to that of this earth. These chemical elements indicate that one of the cosmological constants is of the reality of hydrogen as a universal element.

"So a great leap into the future, into the higher meritorious understanding of evolutionary life is being made. Advanced intelligences from higher dimensions are able to touch the aborigine minds, to explode the many myths and superstitions that have clouded the understanding of man's ancestry in his primitive aborigine past as Homo erectus, to advance an understanding of the mechanical factors of life, which is the germinating nature of the mind.

"So, we have the dichotomy that has clouded the minds of men who have attempted to place upon animal-man the God image with the statement that man is created in the image of God; with only a prehistory in the physical body of 6,000 years, and who did not evolve from the lower animal form. These people have clouded man's mentality by attempting to throttle the evolutionary nature of the electronic system of man!

"It is important then, just as it is to understand the nature of the physical universe, to understand the nature of the physical anatomy, because they are truly bipartisan. The chemical elements which are the matrix for galaxies, suns and planets, and of the plasma field of the universe itself, are the parents of the physical anatomy, One must have a home in which to reside in order to carry out certain edicts for one's self. One cannot function on a busy highway in attempting to conduct his business. In the same sense one must have a body to function through, to direct the business of his evolutionary life. In this respect then we see that the physical anatomy has a great evolutionary history.

"However, the need for a mental body has not yet been determined; nor has it been guessed at by a vast majority of people who have been lulled and mystified as to the what, the why and the how of themselves.

"A science that does not have any purpose or effect in an individual's life, is not a science. In fact, the most important understanding for each and every individual is that he is functioning within an electromagnetic system that is a coefficient of the universal cosmological principle.

"This is the cosmological constant. Each and every individual is buffeted, so to speak, by the pressures of his atmosphere. The electromagnetic lines of force that stem out from a planet and its sun, are in fact directing the individual in some way. It is causing changes in the weather patterns, and changes in the psychological nature and balance of society. The arithmetic averages called the 'Dow Jones' is only the recorded signatures of the minds of individuals who have concerns with certain types of securities, inner securities and outer securities; that is to say, stocks and bonds, which are the basic insecurities of man himself.

"In the same sense, there is a biological movement within the vascular system of the physical anatomy that is registering the pressure of the mental system. So, the electromagnetic plasma, is in actuality, a real and important measurement, not only in determining the real picture of the universe, but of the movement within the mind of man. Man in the long run, in the first and last instance, is the measuring device which is the *alpha and omega* of this cosmic plasma system.

"The universe is infinite because it is created from an infinite creative plasma, if you will, a 'cosmological constant' which is the signature of the *Cosmic Mind,* the father of all life forms! In order for a scientist to become a true scientist, he must remove the last obstacle of the reality of life as an energy system. This energy system must have as its continuity, the energy of higher frequency modes of expression. The present electronic design of third dimensional life contains a sufficiency of examples to validate the frequency and harmonic relationships of all things that are the functioning part of the plasma field that surrounds and interpenetrates it.

"Instead of attempting to resolve the reality of life by experimenting on the trunk of the elephant and determining its length and its breadth, and the chemical elements that make up this sample, would it not be better for that researcher to stand back and take a fuller view of that which he is attempting to sample? Instead of the trunk of the elephant, why does he not take a look at the total configuration of that object? In the same sense, instead of attempting to understand the so-called 'cold or dark matter', why does he not step back, so to speak, and take in the entire configuration of the evidence of an infinite universe?

"The reason he can not do this is because there has been a serious break-down in the evolutionary communication between man and his superman brothers. The reason is that 'hard-nosed' scientists do not want to admit to the reality of the continuity of life from superior forebears, because they will not step down from the pedestal upon which they have gained some supremacy!

"The problem then, of man, is that he has not yet separated himself from his early instincts as an aboriginal ape man, and still is concerned with the survival of himself, in terms of his limited knowledge of his physical environment. It is because of this that he is afraid. He is fearful because he has not seen the full configuration of this environment, and has not yet recognized that his present physical anatomy is the physical appearance of the universe.

"The next step in any scientific development is the recognition of the electromagnetic effect of the mind which is truly the non-atomic electronic reality of life! But of course, this cannot be developed until such individuals give a greater degree of reality to their inner feeling nature, as has been the example of advanced Homo sapiens who have returned from their non-atomic homes to communicate certain evolutionary principles. They are the strong-hearted and the courageous ones; they are the true teachers of progressive evolution of the people who had become tainted by superstition to procure their security, and now are searching for the logic and reason of life's purpose.

"The plasma field principle of the universe, simply stated is this: Life is a plasma field of an interdimensional nature. This plasma field is the matrix of an infinite creative intelligence. It is constantly expanding into higher and greater luminosities. It is constantly contracting, carrying within it the sum and substance for man's evolutionary purpose, a facsimile of the energies of this interdimensional vortical field.

"The picture therefore, of the present physical universe

provides sufficient signatures to give clues to any individual in the forefront of research into the plasma nature and development of galaxies. At the same time, evidence that is developed of the real interdimensional nature of life, and of the physical universe as being the centripetal nature of this expanding interdimensional vortex, should remove all doubt from the mind of man that he is the only human species in the universe. Man is formed of two elements. He is an evolutionary development of centripetal electromagnetic energies that are the geminating force of this atomic dimension, and he is also formed from the centrifugal force of an interdimensional centrifuge - an infinite proposition. The centrifugal force is the mind which is an electromagnetic energy system that contains in its idiom the same evolutionary and engineering principles of the interdimensional vortex.

"To picture one's relationship to a vortex, think of a tornado where there is an expanding matter above the ground and a centripetal or compacted matter below, or almost touching the ground. The physical man is the coagulated matter that appears at the nucleus of the tornado. Physical man was formed from the 109 atoms of the present known atomic elements. But man, the greater and 'real' part of him, is the centrifugal portion that is seen to expand and is infinite in its component attributes. This higher factor of man must be coupled to or joined, for man to leave the lower material frequency of the interdimensional centrifuge.

"What does it take to believe in one's own cosmic nature? What does it take for one to accept that which beats within him day in and day out, every moment of his life, to recognize that this beat is being transmitted by psychic forces; that it is not atomic that it does not vanish after the physical anatomy has returned to its atomic elements. The feeling-self exists, and will ever exist, because it is formed from the higher centrifugal portion of the interdimensional vortex. It is in fact, this higher supernormal part of man that puts to rest all the misrepresentations to the real spiritual construction of Homo sapiens. But first, any thinking man, particularly scientists, who are communicating his findings to the world as a whole, must himself realize the reality of the centrifugal and centripetal actions of plasma fields within the four basic forces of energy.

"This is the evolutionary development of the centrifugal polarity in the alternate movement of infinite, which has been broken down into many fractional modes melded by fractional people who have separated the mind from the body. They in fact, have developed a system that has torn man from limb to limb in attempting to understand what lives within him, and of the whole consensus of the life principle; but they had failed to look into the only means by which one can determine the 'real' nature of life. That 'real' nature is disclosed within the atom itself, because the atom is a television monitor through which is revealed the signal from its transmitter. The atom is simply a television picture. Sometime in the very near future, the furor over the true beginning of the universe will be exploded when new instrumentation will add to, and validate a series of facts that will be complimentary to the cosmological principle of the plasma theory.

"The universe is in fact, a moving plasma that is being contoured from the beginning of so-called time, to develop the present known billions of galaxies, that are being born in some section of the universe. That alone will change the current thought and reduce the imbalanced statements by those individuals who have attempted to form man's understandings with a limited knowledge of his origin and his future developmental purpose!

"It is of course important that the physical basis of life be recognized as an evolutionary principle; but it is only one of two polarities for the individual's evolutionary development. In this way the interdimensional principle of life will be brought to the public through the findings of physicists, who are closer to the *real* interdimensional nature of the life force.

"Such information will free the individual of dogma and superstition. It will do this by lessening his hold upon material life, when his life loses its elasticity at some time at the end of his physical cycle, so that he will be able to see through the 'so-called' tunnel, that will translate him to the 'other side'. That 'other side' is simply the extension of the known physical light-universe to the known non-physical light-universe, where of course the difference is in the frequency of the electromagnetic plasma that is the nature of its dimension. That is the importance of breaking down the superstitions, and the misrepresentation of information recorded by individuals who cannot read their instruments correctly, and by those who simply stand on the shoulders of their predecessors and simply add to the falsity of previously recorded information.

"All in all, we are moving ahead in this time of the close of the present astrophysical cycle, which itself, is a validation that there are changes constantly taking place in the universe. A co-existence between universes will become a reality of conscious life to those aware individuals who have moved through the interconnected corridors between the third and higher dimensions. Near death experiences and past life experiences are examples. The movement of energy as sinusoidal relationships of the electronic and the magnetic effect, and other research findings, will together build a total and complete statement of life, which no longer will be called theory, but which will be a factual and dependable recording of the consciousness of man as an exact evolutionary principle based upon interdimensional physics. Energy does not and cannot live in a vacuum, but is constantly regenerating the sum and substance of itself in whatever form it may take.

"The principles are universal and have been recorded in the distant histories of man but have been forgotten. These universal principles are in fact, the guidelines that must be reerected as signposts for society itself, to reform its very nature. That reformation lies within the individual's prerogative for his own progressive evolution.

"From vast distances in the outermost regions of the universe where uncounted billions of galaxies are wheeling about in their elliptical form, to the planet Earth, where upon its surface billions of people are attempting to equate their minds with the outer contours of their environment, there is a balancing, there is an equilibrium, there is an integration that provides for the peaceful coexistence of the pressures that are formed from the alternate polarities of this great electromagnetic field in which each individual is evolving.

"That interdimensional balance is none other than the realization that each one is a part of a larger whole, and that each individual has an important purpose in validating his consciousness. One who sees his relationship within the great *infinite plasma field*, will find himself in service to mankind!

"We leave you with this thought that philosophies and sciences that attempt to explain life by separating the parts from the whole, can only do injustice to that necessary integration of mind and body that all individuals so sorely desire. Right in your own garden, in your own back yard you have all that you need to learn of infinity. By giving of 'Self' do you begin that alternate movement, which becomes an integral part of the movement of one's physical and mental nature. As the two forces begin to equate with each other, you will find that your physical life becomes healthier and more balanced as you have equated your true inner force and nature with the tremendous down-pours from the centrifugal portions of the interdimensional vortex - the *cosmic fountainhead,* and into the receptors of your mind. The *science of life* is nothing else than an understanding of the four forces of your own nature.

Chapter 5

ATOMS, MAN AND THE UNIVERSE

"The revolution in physics which has already made ripples on the surface of the scientific community over the past eighty years, is not the entire completion of developments which were antecedents of this revolution. We must remember that man is a star traveler who contains within himself the answers to inner and outer space.

"There are those who are traveling throughout space and time who have, shall we say, constructed a spacecraft that contains instrumentation that is far more in advance than other scientists, who are involved in this search for life in so-called outer space. The search for extraterrestrial intelligence is the basic ongoing factor indicating the great increase in interest in the realization by those individuals that indeed man is not a singular creation of the vast infinite but must exist on extraterrestrial planets which are seen to be the spawning grounds of intelligent life in exterior space.

"The cosmos is therefore the universe. The universe is the reflected picture of an incomprehensible infinite, expanding in such a fashion that the basic rules of third dimensional science are no longer operable. Amongst scientists are those who have constructed instrumentation that provides them with a greater ability to view the cosmos and to recognize the physics in which this movement is making life recognizable to those individuals who are living within that spaceship; for they do have a greater comprehension and therefore are capable of moving to distant planets.

"These spaceships, of course, are planets themselves! Planet Earth is, to all intents and purposes, a spaceship because it is moving through space in five different directions. Albert Einstein was unwilling to accept the developmental impact of the theory of quantum mechanics to explain the entire nature of the movement of energy. Einstein indicated that the quantum theory by itself is not complete, that there are other principles of energy that must be formulated before there can be the ability to arrive at that completion, so desirous by physicists; a unified field theory where gravity or the graviton, the electromagnetic force and the strong and the weak force are all united.

"For that unification to take place, scientists will have to become open and aware of a factor which is still not an accepted feature of mathematical theory. This is the realization that man himself, is the subject and the object which he is ascertaining as his exterior world, which cannot be examined without his own subjective involvement. The very involvement of the observer changes the observable fact. Therefore, there is in this respect, the contrast between those physicists, such as has been with Neils Bohr and Einstein who contradicted each other with respect to the means through which the field theory of energy would be integrated within one complete mathematical theory or mathematical statement!

"Einstein's point of view was not accepted as he indicated, and as we have also stated and will restate that, "God does not play dice with nature"; meaning that the minutest subatomic particle obeys the same basic principle in which the macrocosm functions. What that principle is, is for future physicists to discover. Mathematical physicists who are presently working with these realizations are aware that the mathematics of linear measurement is not capable of providing a true picture of energy as it functions in another configuration. This indicates that the topological mathematics is on course and more capable of giving a true picture of the reality that exists within the expandable movement of energy.

"This realization or factor has been validated using the basic measurement of the 'red shift' and the 'Doppler effect'. It is just on this basic premise that there is an enigma and a contradiction because man is the subject, and he is attempting to validate what he considers to be the object. But as man is entwined with the object, he cannot devise instrumentation that will provide him with the reality of the picture that he is attempting to gain with what he considers to be objective measurements. It is on this basic understanding that we could point out that the third dimension is basically a mirror of the fourth dimension.

"Certain scientists have used the term, 'Looking Glass Universe'; and they are close to the truth. For indeed, the universe that is the third dimensional proposition of life is a mirror which is reflecting the objective nature of life. That objective nature, that objective feature of life is the fourth dimension. The third dimension is basically, and to all intents and purposes, the video screen of life, formed by the demodulation of the higher frequency energy wave forces stemming from the fourth dimension into the third dimension screen of life which is man's reality, the atomic picture that has been so refracted upon the individual's consciousness.

"To all intents and purposes, it can be seen that life is very similar in all respects to a television receiver. What has been missing has been the realization that the entire intent of life has been projected from another dimension. The quandary, the enigma that has pursued science as long as it has been known in the last several hundreds of years has been just this factor; the deployment of instrumentation to indicate the nature of all factors of life within the frequency of the light, and the divisible factors of the atomic spectrum. But the understanding of the transmittal of energy, the crossing of the barrier reef, to all intents and purposes, from the fourth dimension to the third dimension is, in every respect, something of a no-man's land. The accepted features of atomic life are no longer recognizable!

"It is this very factor that is the present debacle of scientists, because it is their determination to understand the reality of the atom; its creation and its mechanics and function, with which elementary particle physicists are so concerned. Subatomic physics and the quantum mechanical view is *no-man's land,* where time and space no longer apply as space/time is a demodulation of higher factors of energy.

"These higher factors of energy are no longer separate in their movement from polarity to polarity, from positive to negative, charge and discharge, and are functioning now in a different environment. There is no measurement that can now determine the movement of this energy using the measuring sticks of third dimensional physics. This is the basic enigma that is facing particle physicists, and indicating to the entire scientific community that life is an ongoing process in which man is not what he thought he was. Man is therefore not a mechanical instrument that obeys the laws of nature, as long as nature is considered to be only a third dimensional pro-position!

"The physical body, therefore, is not the complete definition of man. The physical body is itself dependent upon and indeed must function within the laws of third dimensional physics as momentum, inertia and gravity, and the anatomy may be damaged when any of these three factors of energy are not considered. Should a man jump from the top of a tall building, he will suffer the effects of all three of these basic physical movements of energy. The momentum will carry him to the street below. The inertia caused by impact of the falling body and the occasion of the constant downward force, so to speak, activated by and through gravity, will cause the serious damage to that physical anatomy. These are principles of energy that must be recognized by all mankind. However, man who is living within this atomic body is not constructed from physical atoms.

"The construction of the energy which is the true consciousness of the individual, obeys other laws. It is these other laws which as principles, are the determining factors of energy in different dimensions of higher frequencies. It can be seen then that man's mind can move through walls, planets, solar systems, galaxies and universes, as well as through third dimensional states of matter.

"The mind of man is constructed on interdimensional factors of energy which has its home, so to speak, on higher levels of existence. The mind of man is a statement of a force that is constructive and has been developed by each and every individual so living in a physical anatomy or in a non-physical anatomy. That mind is the truth of the statement that *life is interdimensional.* The mind is the objective nature of a field of force which can more properly be called an anatomy. It is an energy body that exists in the same sense in which third dimensional matter exists because it is receiving its knowledge, its intelligence, and all of its necessary respondent direction from a transmitter which is never shut down, but is constantly projecting information to its receiver providing whatever intelligence is being projected from the transmitter.

"Man, therefore, is no different than atoms, planets, solar systems and the larger universe. Man has a transmitter that functions from a fourth dimension; that is to say, a non-material dimension. It is this *non-material dimension* which has not been given acceptance by third dimensional physicists, because they have not been capable of ascertaining that the signals for the reflection of life on third dimensional planets, were received from another source - a source which did not originate on the terrestrial planets; a source that is not dependent upon the atomic model. It is not dependent upon the movement of energy in the third dimension. It is not dependent on man made scales of time. This transmitter is a completely and wholly integrated energy form.

"An individual who is completely involved in determining the truth of the mirage of space/time, must himself, wake up to the knowledge that he or she, must recognize the validity of his psychic abilities in any interpretation of data from instrumentation. Because those abilities which are paranormal, existing within all human beings, are the sense structures of his fourth dimensional body and will validate for that researcher that he is standing with his two legs astride, one in the third dimension and one in the fourth dimension!

"In every respect Homo sapiens, as he is developing his paranormal abilities, becomes that much more alert to the realization that at all times his mind is functioning from his fourth dimensional home and his body is receiving the signals of this fourth dimension transmitter. This is the *waking up period,* therefore, for those individuals who are considered to be the detectives of energy and who are attempting to run the trail down and to validate the nature of energy through the system of science, using logic and reason. They wish to resolve once and for all, the truth of the extended nature of life, not within the narrow confines of living from child to old age and then shedding the physical anatomy at death, but with the knowledge that the continuity of energy itself is a scientific statement, that man cannot stop in his analysis after the completion of the life of an individual in a physical body.

"That is to say that the death of the atomic body or its return to the elements of its atomic world, *does not remove* the energy which is the *consciousness* of the individual from its impaction into the greater energy field. This means, to all intents and purposes, that any science to be truly valid must be concerned with the nature of man's consciousness as a continuity which bridges the gap from the physical reflection of life to the *non-physical reflection of life*. Thereby is the true intent and purpose of man resolved through this recognition, that it is the mind of man that is carrying the true information from an integrated world - timeless and spaceless!

"A *unified state of consciousness* and a *unified field theory* are one and the same thing. To develop the understanding of the unified field theory, one must have a unified consciousness which takes into consideration the fact that man is a dual factor of energy. The picture of the cat chasing its tail is a metaphor describing certain research techniques of scientists who are using obsolete means to resolve the dichotomy of life with the desire to find one complete principle that would explain all the anomalies of third dimensional life.

"The anomalies are only in the mind of the individual

researcher. They do not lie within the all-embracing movement of energy. The infinite is completely unified within its own movement and the movement of man, which is his developing state of consciousness. Or, we can say that it is the instrumentation developing within man, which is his mentality, that has not been capable of viewing the polarity movement of energy. He sees one movement of the electron and then he sees it on the opposite end of his observation. But the electron is the extrapolation of a wave form functioning in two polarities. Yet when he attempts to make an objective statement as to its makeup, he does not have a complete picture of the electron, or any other particle for that matter, because the particle is not in one singular space.

"This is the evidential statement for the need for a new mathematics to be able to relate the movement of the electron and to give it a concise and definite description. That definitive description cannot be made using third dimensional principles of physics. This is the basic problem! Einstein, Max Planck, Schrodinger and others have been the pioneers, who were moving in that direction to explain the dichotomy of the subject and the object, and were providing in their awareness, the beginning of a new mathematics.

"A new mathematics basically takes into consideration the understanding that man himself is energy and in his own association to the electromagnetic field, is changing the nature of his observation. Basically those scientists who have realized the dichotomy that lies within the nature of their own consciousness, have accepted the nature of their quest for the unification of life as it could be measured in a mathematical manner. They have realized that each and every individual, no matter what his trade, profession, or his supposed intelligence, is a developing factor of a greater infinite; that he cannot resolve the dichotomy of life using the set square and the microscope, as long as he is not capable of measuring his own pulse and the nature of his own heartbeat!

"The greater the researcher who is, in every respect, a measuring device; the greater his understanding of his own measuring device, his own electronic energy system; then the clearer will be his ability to reflect into the greater cosmos and to make concise statements as to the nature of reality. Science is no different than any other developing factor of life in which man is learning of the nature of his own involvement in the schematic of evolution. Scientists themselves are witness to the fact that they have been involved in the detrimental and the divisive nature of society. Yet they have not known of the dangers to which they had lent their knowledge and their names, by taking sides on issues having to do with national pride or in the desire to advance the ego.

"Scientists are human beings and scientists must, in their recognition of this, realize that in the desire to ascertain the nature of reality using various measuring devices, whether they be in mathematics or in electron microscopes, the object of this quest must be selfless rather than selfish. In this respect those individuals who were able to let go of personal factors associated with their own opinions were more capable of being *true to themselves.* With no consideration and concern that they were out of tune with their colleagues, they held fast to their own inner knowing, even though the basic means for proving this knowledge through instrumentation of third dimensional means was not available.

"Einstein's theory was a proof of his communication with more advanced intelligent people who live in non-atomic worlds and from which all valid information is received into the third dimension. Albert Einstein received a fourth dimensional picture of the movement of energy and therefore, within the boundary lines of the present mathematics, was able to develop the necessary symbology which could be an accepted factor within the present 20th century. $E = MC^2$ is his famous algebraic statement of energy.

"All third dimensional life is the result of radiant light projected from the sun, which is itself the property of a higher energy field. Basically, this was the proposition that Einstein was stating, that energy is cycular in form.

"If one had the paranormal abilities as Einstein indicated to a certain degree, expansion of the universe would change the flux fields of energy in which the speed of light would expand to greater frequencies. That expansion unit would then be the nature of the continuity of energy; and the continuity of this energy would resolve itself into the expansion of the movement of energy into higher fields whose frequencies are beyond the atomic element or beyond the speed of the third dimension light band.

"Light is therefore not fixed but is constantly expanding. To that extent, the planets that function in these higher energy fields are planets which in themselves contain a greater measure of intelligence. Therefore, man is himself an expanding factor of this movement of energy; and as he leaves his physical anatomy, he takes up his position, frequency-wise, in a particular energy body that has been so developed or evolved whereby he continues with his life as he was living it through the denser or more impacted frequency energy wave forms of a physical body. The energy body which is non-atomic, then becomes his integrated body and through which he is able to see out into a world that is not made from the lower or denser energy wave forms known as atoms. The atoms are isotopic in their higher frequency formation.

"Many scientists in different fields of enquiry have ascertained and conjectured as to the continuity of life from the physical consciousness to the spiritual consciousness or, if you wish, to the non-atomic and more clear consciousness in a dimension where the strictures of an atomic world are non-existent. Many of these individuals have been inscribed in the Halls of Fame as intellects who have added to the positiveness of life, and as humanists have, in many respects, laid down directional signs for man to follow. They are the forerunners who have laid down these directional signs in various disciplines of life. It is the present 20th century in which the basis for progressive life has been stabilized in the direction which we have been relating; a direction in which life can be seen to be purposeful and continuous, rather than singular and incomplete and where the signposts themselves have been demodulated into a variety of language systems.

"The unfortunate factor is that many of these language systems with which individuals could determine a positive direction, have been changed. Many of them have rusted away and have been corroded. Much of this has been in the history of past civilizations that have lost their identity and recognition by the civilization that continued after them. These civilizations such as Atlantis have contained within them the basic corrosive factors which have made the present incomplete, in the minds of those individuals who are supposedly opening up the logic and reason of life.

"These very individuals are those very self-same people who formerly lived in these ruined civilizations, and had lent their scientific knowledge to the structure of those societies. Since all people are continuities from previous civilizations, there is not one man, woman or child who has not lived in the Roman, Grecian or Egyptian civilizations, or previous civilizations!

"It is in the previous civilization of Atlantis that much of the corrosion and much of the amnesia of today can be explained. The reason why certain scientists, using basic similar measurement devices come up with different interpretations is explained by their association with the negative nature of science used in the deployment of its power, twelve to fifteen thousand years ago. These individuals are themselves unaware of their past lifetimes; and in their unawareness are carrying the very nature of the corrosive and more singular factors of their knowledge into their present 20th century. In this amnesia there is a great psychological pressure - a pressure in all respects of insecurity and guilt!

"To the extent that these individuals are aware of their continuity from the past, can they be freer to open up to the new physics and to recognize that the only block that holds them from learning the *New Science*, is their own mental amnesia; that the nature of their acts in the abuse of energy, has been a continuity in 1945 with the detonation of the first atomic bomb. Scientists who were involved in the construction of the first atomic weapon are continuities from Atlantis and, of course, previous civilizations.

"These physicists and engineers were carrying the basic proposition of their understanding of energy to the point where it was recreated for purposes of national interest and loyalty to the country in which they were citizens. However, the loyalty they demonstrated was not to the truth of their humanity in their evolutionary development. Their loyalty was therefore a continuity of the acts in which they were so involved from previous civilizations. Until the very act of the detonation of the atomic bomb was the full realization brought to their consciousness of the truth and responsibility for their actions. In this respect these individuals are aware of their guilt and the great impact that it made in their consciousness.

"This is the knowledge that must be a part of all individuals who are involved in the advancement of science; in the advancement of the knowledge of evolutionary life. They can, in this respect, with their realization of the true principle of science, of cause and effect, know that holding back the constructive, positive development of science would only be a reactionary device to hold on to the security of some past-life remembrance. In many respects these past-life remembrances are not complete in that they do not take into consideration the involvement of that individual within the fullness of the act which had destroyed the lives of millions of people and would not readily be acceptable in the mind of that individual.

"So it can be seen that science is not truly science unless it involves the understanding of the individual in his life-to-life continuity from his past and, of course, into the future which is the regeneration of the past. Therefore, those individual scientists who are presently involved in their past involvements by holding on to obsolete means of measurement are, in their own manner, intractable and have not gained the relative understanding of their own infinite property. They do not yet know that they are not material elements to explode and to create through the bondage of some cement or atomic materials but rather, that they are and ever will be the material that is the stuff of infinite intelligence that cannot remain bonded in one place and expect to expand in its desire to live!

"The New Science fully embraces the basic truth of life as an interrelationship between two polarities. The atomic mirror of life is one polarity seen through the resolution of a physical anatomy and all factors of terrestrial life. Man is also a resolution of another polarity that is the positive element of this dual construct of the Infinite. This positive polarity is mental, since it is nonatomic, and has as its base an energy structure that is fourth dimensional, timeless and spaceless. It is this energy structure that is the nature of the reality of man.

"That energy structure is his electromagnetic body, an instrument which functions in-phase with the higher frequency energies existing in the fourth and higher dimensions. This is the unified field that is so highly being quested! It is already within the embrace of any individual for him to discern, but it must be applied. It is in this application that science is now beginning to make a great stride forward. Science, as it is a continuity from the past endeavors of man to learn of the coexistent dimensions, the third and fourth, is also in many respects, a victim. At the same time those individuals who call themselves scientists are victimized by the lack of knowledge of their past!

"Therefore, scientists as a whole, who work with the apparent contradictions of life for a greater understanding of the evolutionary design of life, must themselves become students of this evolutionary design and accept within the framework of their own work in the laboratory the fact that they are the *past-life designs* which are inflicting a certain degree of discrimination and incrimination into their work. Until they have become aware that they are fourth dimension constructs, as had Albert Einstein, Erwin Schrodinger, Boltzmann and others, they cannot possibly advance the disciplines of science. They will only be carrying the same imagery from the decayed civilizations from which they originated their present work.

"This is the time when the bells ring out and the clarion sounds call to man to wake up to the existence of his true reality! It has been a reality that has been with him but had been undisclosed because of the over-population of material features of life which have, in every respect, fogged his higher sense device that is his mind which has access to the transmission signaling the truth of his reality. As these overhanging weeds are removed through the logic and reason of the evidence of the continuity of consciousness, man will realize, through the help of science, that he has always been functioning from the lateral higher frequency dimension in order to express in the lower physical dimension.

"The constructs, therefore, of scientists in whatever fields he is functioning, is the means by which a cohesive description of the nature of evolution and of man's continuity into worlds of a higher design can be furthered. Let us realize that since the infinite functions in a dual fashion, this duality is constantly receiving intelligence from the positive pole, the mental side of life. For as man is living through his physical anatomy on the negative pole and as a student of life, he is being given information of the evidence of his *Teacher's* existence. In every respect then, the *Teachers of man* always have existed in a higher mental dimension!

"The Teachers of man are Brothers who have long ago learned of their unified state of consciousness, and of the unified field equation. They have learned of the integration of the sine wave's alternate polarities; and they have realized and have been communicating this knowledge of the reality of infinity, of the higher Intelligence that is the constant impetus of life to all people who live in whatever dimension it may be. In the constant movement and deployment of life in the lower terrestrial dimensions, individuals as teachers, move amongst their fellowman in different areas of life. These are individuals who have taken on many names, and are in every respect, Brothers who have attained to that awareness of the inner and outer nature of life as it becomes a reality in each dimension.

"In the third dimension this reality has become the maintenance of the physical body, a reality of life that is regulated by the radiant field of light. As man's consciousness develops into the wider vortical sweep of infinity, within an expanded centrifugal field force, he becomes aware of another reality, and thus develops certain properties of intelligence which expand consciousness. The reality of man is relative to the dimension in which he lives and the consciousness he attains.

"Those individuals whom we have introduced to you as Brothers, are now awake to an expanded *reality of the infinite.* They see themselves and act accordingly as participants within the movement of life; and in this movement are the hundreds of trillions and countless more billions of peoples who are all functioning as electrons are moving in their trajectories to expand the nature of the infinite.

"Subatomic physics to a great degree will reveal that mankind, as he lives in his societies, is functioning in many respects in the same or similar manner. The principle by which the infinite functions in its finite derivations will reveal the great sine wave principle of life; and in this revealing will man wake up to the understanding that all of the anomalies and injustices, so-called enigmas and other separatistic features of his physical life are based on his own mental inability to see the duality of himself!

"Subatomic particles behave in the same manner as does man in his society; and yet in the seemingly random nature of the movement of these subatomic particles there can be seen the underlying principle which affects these random movements and which joins them. And that is the same factor within societies of man that seem to function in a random fashion but underlying man is the basic correlative and integrative principle that man individually and as a whole, is a unified relationship of a greater whole!

"This greater whole is the composite factors of all finite properties which make up all the elements of third dimensional, fourth, and higher dimensional factors of life. Therefore, quantum mechanics can be seen to be the means by which the integrated nature of energy is functioning. In other words, evolution will be seen to be the basic principle, not only in the recognition of the astronomical cosmological mechanics of life but the individual man, the participle man, the energy quotient that is man, will be seen to be functioning in the exact pattern as is the movement of star systems in their expanded form!

"Dependence upon instrumentation that has been constructed through the mathematical symbols of Cartesian mathematics will not suffice to explain the futuristic mathematics of topology and other developments indicating the curvature of space. The nature and movement of star systems in a direction that indicates the expansion of the universe, is an infinite proposition. Although the third dimension can be seen to consist of an expanded nature, it is also a closed universe to the extent that it takes on an entirely different picture once the speed of light is resolved into another dimension.

"Hence it is that the Special theory, so propounded by Einstein, and the quantum principles that have been so propounded by Schrodinger and Planck, are not necessarily different because there is a matrix in these two perspectives of the movement of energy. It is in the resolution of these concepts that the enigma of energy will be solved as to the completion of the basic principle of the unified field theory, a principle which will explain the motion of the weak and the strong force, the gravitational and electromagnetic force.

"Both the quantum mechanics of Planck and the evolutionary theory of Einstein contain the basis for the resolution of this question. But the enigma is within the constructs of the individual researcher to add the final note. And it will be through one of those individuals who, as a Brother who had attained to an awareness of the continuity principle of energy in its dual relationship, from the third to the fourth dimension, carrying both positive and negative signals that will demonstrate the resolution and provide the answer to this problem. That will be the dawning of a New Age of Science!"

Chapter 6

ENERGY AND MATTER, A FIELD CONCEPT

"The mathematics of small numbers, of small units, of minute particles, is the thrust forward that is presently being engaged in by physicists who are involved in the constant development of greater and greater particle accelerators. The physics that is being researched, is hopefully the movement of energy from the macrocosmic world of the five senses into the sub-microcosmic world which does not obey the synonymous factors of the classical laws of physics of momentum, inertia, and gravity.

"Individuals and society are studied by sociologists and psychologists in terms of integration and lack of integration, interrelationships and movements within the environment of a family or an industry in which individuals are functioning around a central core, extending out into the nation and to the planet as a whole. We can see that atomic particles seemingly moving in anarchical manners have a similar relationship to Homo sapiens as he is a function of many complex orders and that these complex movements are themselves related to a base-plane frequency.

"The base-plane frequency that is operative within the sub-microscopic level has a continuity and a relationship from which the smaller becomes the larger or the larger becomes the smaller. The centripetal actions of the vortical energies functioning sinusoidally through the electromagnetic lines of force are the isochronisms which give the stability to the vortex. The movement of the energy wave forms in this vortex, as an example, are moving in two directions at the same time.

"Now this may seem rather an anomaly, but energy moves in many different relationships unknown yet to the instrumentation of a third dimensional physics.

"The interesting factor is that the vortex that we are describing is in actual resonance with each human being, formed from hundreds of millions of vortexes within his electronic anatomy, truly, the transmitter that makes possible all known third dimensional features of life. Of course, the highest evolutionary development of third dimensional life is man's physical anatomy. The design activating the electrical impulses that move the complex component parts of the physical anatomy are derived from the brain, a computer transforming device which contains the signaling system, making possible the continuity of man from his fourth dimensional anatomy into a third dimensional form, the physical or atom body!

"Particle physicists are attempting to understand the subatomic forces of energy by colliding an atom with electrons to explain the paradox of the atom's inner structure and movement!

"Of course, this is not possible! If one should separate an individual into arms, legs, feet, fingers, torso, head, you would have a corpse of divisible properties, but you would not have a man! If you should then take this corpse and do an autopsy and continue to separate all the organs, you would then have only an atomical accumulation of cellular structures, but you would not have a human being. You would not have a living, breathing force that is receiving its intelligence from its only possible derivation, from the transmitter that gave it its life, the infinite creative intelligence!

"The transmitter, since it is not of an atomic formation, does not contain cellular structures, organs such as a heart, liver, brain, lungs, etc. The transmitter is an electronic energy system, functioning as a carrier wave from its interdimensional design, demodulated into the nuclear dimension. It is this nuclear movement within the centripetal structure which explains all factors of life on the third dimensional spectrum of atomic life!

"Such frequency structures which contains the nuclear form of the compacted intelligence of the vortex, are now propelled outwardly as a lower dimensional structure where all factors of intelligence are now reversed. What was right is left, positive is negative. The polarities that were the interactive, integrated nature of the energy wave forms that make up the sinusoidal wave forms of the vortex, are now reversed. This is the great mystery in the reversal of the movement of energy and in its separate relationship to its previously experienced unified form.

"The fourth dimension is called such because it contains the unified field of the four forces of energy that are separated in their third dimensional equivalents. In the unified field, energy moves in a cycular manner and therefore, time and space do not exist separately. The weak force and the strong force govern the interior movement of the atom, while the electromagnetic force and the gravity force govern the exterior movement of the atom. These four forces exist in a stable relationship in the higher dimensions because they receive their information from the unified nature of their integrated structure!

"Philosophers, physicists, chemists, and biologists, etc., are examining the basic properties of the atom. Medical doctors and surgeons examine the individual organs of a physical anatomy to inquire into the causal factors of some disease that has resulted in the death of an individual whose body is now laid out on the slab in a dissecting room.

"However, more correctly, that body no longer belongs to him because that individual who was the electronic force, is no longer occupying this third dimensional receiver - and that is factually what it is. He has long left the scene of whatever has caused his death because the receiver cannot function without a transmitter. To all intents and purposes, the television receiver has been shut down because the plug was pulled from the connection that gave it access to the transmitter. The programs are no longer capable of being viewed on the face of the television's phosphorous screen. This is the nature of the physical anatomy; and to a great extent, is the research tool of man to understand the physics of his transmitter!

"It is truly the transmitter that is the most important factor in the ongoing research of any individual who is alive through his contact with his fourth dimensional self! The fact of the matter is, that the majority of people are not aware of the proposition, that they are, and have always been receiving interdimensional frequencies of a high order which have made possible the stabilization of their life through the lower frequency atomic body!

"Evidence exists to indicate that the termination of the physical body does not necessarily remove the evidence of the existence of the *Transmitter.* In the history of the paranormal, evidence has been established of the existence of the extended nature of man's consciousness. That extended nature provides the clue that man's mind is a separate factor of his brain. Yet of course, the mind is related to the brain as an oscillating in-phase circuitry, demodulating the higher frequencies of the carrier wave into the receiver.

"As that system evolves, it develops into an intelligence

that we can call 'mental'. A person who uses his mentality is therefore functioning in a trilateral manner, encompassing information that he has garnered from past lifetimes, which can be considered to be the accumulation of life experience stored in the subconscious.

"The conscious mind is man's present association of information as it provides him with the recognizable features of third dimensional life. But the intelligence that truly gives man the appearance of integration is the association of all factors of his life that have been accelerated and digested from life to life, in hundreds of thousands of life experiences, polarized from the superconscious.

"Man cannot be intelligent until he realizes that the physical body is simply a receiver and that it cannot provide new concepts, but can only solidify understandings that he had previously experienced from his past. However he has the capability to improve the reality of his Consciousness with the realization that he can live minus the physical anatomy; that life after the death of the physical body, is a functioning reality! In other words, he is aware of the *interdimensional nature of his consciousness.*

"The brain, therefore, is a complex machine that reflects the intelligence of it's designer. The information accessed in this research, will provide the researcher - the neurobiologist and the neuropsychiatrist, the deeper understanding that this complicated device, which is a transceiver, did not develop in a vacuum! For example, the growth of the fetus in the womb of the mother, indicates the great intelligence that is interacting with the cells of both male and female parents.

"The mitosis of the cells demonstrate that this great intelligence knows specifically the characteristics that will be the completion of the growth or the development of these genetic structures. In other words, the very act of birth itself and the continuous regeneration within the movement of the cellular structures, of Homo sapiens, animal, plant, vegetable, or mineral, is an indication that a *Force* is being extended from another dimension. In this way an intelligent energy system is revealing its presence as the receiver. In this manner the newborn child is recreating the specific molecular structures which will mature into the human physical anatomy.

"This Force is as a transmitter communicating to man his basic architectural design, his purpose in life and his need to understand the nature of his present progressive development. It is this knowledge that will explain all inquiries into the reality of an infinite creative intelligence, enfolded within his pattern of life. The physical anatomy is the evidence that life, transmitted from higher non-atomic dimensions into the present atomic dimension, contains the imprint of a program that is the next step in the evolutionary development of the species man.

"As man learns of the infinite intelligence that lives within him, he will validate a physics of life which is interdimensional. The basic understanding necessary for the man of science is to see himself as a factor of his own research into the atomic particles that he believes will explain the field force of the atom. Man cannot separate himself from the movement of the interdimensional factor of the infinite. It is in this separation that anomalies are witnessed.

"Scientists have gone as far as possible in the use of measuring devices developed for the understanding of the factors of the mind. Yet these discoveries are still dependent upon the rudimentary measuring devices which have used the properties of linear measurement.

"Initially we proposed and made the statement that energy is interdimensional. The cosmic egg, so to speak, that unknowable, is Infinite Intelligence. The effects of the movement of this interdimensional energy are the recognizable features of present third dimensional energy. However, one must function from the bottom up and learn of these interdimensional principles as they are interrelated in the third dimensional energy system of the physical universe.

"Present work currently being researched in particle physics and cosmology is moving in the direction of a greater understanding of the expanding features of the universe. Rudimentary measuring devices, cannot function to explain present theories of the *Superforce,* and the *Cosmological Constant.* Present mathematics can no longer work to apply realistic answers to the movement of the energy of this Superforce!

"The third dimension is a mirror effect of the fourth dimension! All matter therefore, that contains the basis of what can be considered the solidity of life, is a reversal. Just as the lens captures the light of some object outside of itself and brings it to a focus and impinges the property of that object onto a sensitized plate as a polarity reversal, the complex instrument of man's mind is seeing in all respects, upside down and inside out. But because of additional factors that have been programmed into man's consciousness throughout his long history on earth planets, this dichotomy has been reconstructed in what is seen to be the present features of life in all relationships, ie., as he walks down the street or sits in his chair, or watches his television, or travels in his automobile, his airplanes or trains. He is, in actuality, seeing in reverse, and he has more or less learned how to automatically reverse the polarities so that the objective world gives him a feeling of integration or wholeness.

"In this respect, the measurement of the expanding nature of the universe, as galaxies move out at tremendous speeds away from each other, is explained through third dimensional means of measurement, the spectral evidence of radiant light in different frequencies as measured on earth! But since he is living in a reverse world, these measuring devices are no longer capable of providing intelligent information as to what actually is taking place.

"Therefore, the evidence for the big bang theory for the ostensible beginning of the universe as a great explosion of a primordial atom is the measurement of a mirror reflection and not the evidence of the original or first energy. In other words, the causal field of a higher dimension that sustains the third dimension has not been recognized. It has always been true that man, in his attempt to know about the unknowable, has been chasing the tail of the cat and can never catch up with it or catch on to it because he is, in that anachronism, constantly running away from the answer!

"If he would only stand still and do some intelligent reflecting, using logic and reason, as we have indicated, he would see that the fetus is the result of the wedding of two eggs, male and female, which results in the very complex association system of cell structures that culminate in a physical anatomy. The most complex design of life is man himself. The very activation of such a genetic development is an indication of a *Transmitter* responsible for the resultant picture that unfolds before the very eyes of a scientist or any individual who is thus watching the birth of a human being.

"Hence, in whatever field of research or whatever the discipline of a scientist, the scientist himself must become his own research. He must do research into the nature of his own thinking processes. And since most of the thinking processes of man are reactionary, because he is dealing with the primordial forces of his own past, these are the forces that must be acknowledged to exist. These are the explosive forces that live within all mankind; and these are the forces which are presently causing untold misery amongst billions of people on earth!

"Chemical dependence is one factor indicating the explosive nature of certain negative factors of life that have accumulated within the mind of any one individual. These negative energies are exploded when the individual's energy field is imploded by the presence of his malformed sine wave structures, resulting in an out-of-phase alignment of the individual's mental intelligence.

"Such energy in all respects, is the dichotomy of an individual who is not in complete awareness of the interrelationship between his physical life and his mental life, and where the physical desires of the atomic body, as a mechanism for sensing life, has gone wild. Such an individual is pressing into his physical anatomy additional volatile energy that is building to such a pressure that any one incident could ignite the voltaic energy to implode.

"This is exactly what was taking place when the atomic bomb was exploded in 1945 over the city of Hiroshima in Japan. The fourth dimension exploded through the atom killing and injuring over one hundred thousand persons. The nuclear force of the atom was the gateway to the protonic force containing the compacted energy of its higher dimensional intelligence, a fourth dimensional force that was unintelligently directed into the third dimension.

"That fourth dimensional force had been playing its part

in an intelligent fashion through the eye of the atom, to achieve some specific purpose in the third dimension. However, forcing that atom to lose its continuity with the third dimension, loosened its continuity as an intelligent receiver and allowed the pressure that was activating it to explode unintelligently, thus causing negative consequences in the third dimensional field. The destructive force of the atomic bomb is caused by the eruption of the unified field force of the atom's fourth dimensional intelligence into the third dimension.

"That is also the consequences of the energy that is the functioning factor of each individual whereby the understanding of the physics of man as an evolutionary design and of his mind as the fourth dimensional factor, gives him the ability to function in the atomic world. But when energies which are at all times being projected from the higher frequency dimension are allowed to lose their direction, we have a breakdown within the developing force of any individual man.

"That directive force is the necessary feature of individuals who are attempting to further the reality of the interdimensional design of life through the development of their own understanding in the fields of knowledge which are the means of adding to the recognizable design of man. Hence, the analogy between the atomic particles that are separated from various other atomic elements and examined to give evidence of their relationship in their movement, purportedly giving a personality profile to the atom and of its field force, thus providing an understanding of the grand design of life from its sub-microscopic form!

"In the same manner, astronomers are attempting to understand the nature of the universe by examining the movement and mass of planets, suns, and the galactic formations in the universe. Cosmologists are more concerned in attempting to understand the entirety of the macrocosm to provide a coherent understanding that would explain the creation of the universe. This is the pushing drive of mankind as each and every individual known generically as Homo sapiens is a product of the cosmic egg, the infinite creative intelligence!

"Yet there are many peoples presently wearing the robes of high office, who carry with them great out-of-phase energies which are possible atom bombs! They are carrying explosive energies from past lifetimes and, unknowingly or knowingly, are sitting on a keg of dynamite. But more particularly, they are sitting on the explosive possibilities of causing not only their own destruction but the destruction of many hundreds of millions of people because of their present ignorance as to the nature of their own energy structures, formed from past negative life experiences!

"The mind is therefore, without a doubt, the greatest demonstration of the existence of a *higher creative force.* This higher creative force is, actually and in substance, a property of all mankind! That higher creative force has been so recreated as a computer to provide for the functioning ability of Homo sapiens to live, for it is constantly being rectified and energized by a still higher Force. This is the basis of present scientific development whereby those individuals who are the products of the higher mentality and who have gained the understanding of the physics of the interdimensional cosmos, are presently motivating other persons with their wisdom. These intelligences are the creative forces that must be recognized by the teachers and the leaders and the officers of government and industry, who are responsible for the order of life within the framework of society lived by mankind in the various parts of his planet.

"But if life exists on planet Earth, one small planet in one solar system of one galaxy containing over two hundred million suns, then the billions of galaxies containing the movement of intelligent energy, must therefore contain the intelligence of hundreds of billions of humanity! And it is this evidence that is slowly but surely coming to the minds of those who are the researchers, the advanced scouts who are the physicists, the astronomers, the cosmologists, and anthropologists. And all peoples who are thus looking into the nature of energy, into the protoplasm of cell structure, the protoplasm of the Earth or a sun, are waking up to the fact that the ongoing process of life is indeed awesome and has always existed and did not develop in the instantaneous explosion of a primordial atom.

"The primitiveness of the big bang cosmic theory for the beginning of life indicates the ignorance for the logic and reason of life that exists within each individual. No individual can consider that he will not live beyond the last tremors of a physical anatomy. There are few individuals who feel that they are other than their physical body. Those who do so are yet young in their evolutionary development in the learning of their participation in this great melody of life. They have not yet awakened to the recognition of the stream of consciousness of the mind, a complex structure of a fourth dimensional anatomy which is the electronic or psychic anatomy of man.

"But those who number in the thousands, are the prime movers, who are digging the necessary postholes into the features of intelligent life on many planets. Their spiritual beacons are the integrated design of a light that is reflected in a 360 degree circle. It is this *light* that has been made visible within the physical features of a third dimension.

"This light is the evidence of advanced, intelligent scientists, philosophers, educators, and people of a higher framework of like understanding, who are functioning in a dual fashion, having crossed the horizon, invisible to physical man. But they have made their presence visible because of the erection of a great beacon which shines to provide direction for those travelers into the stars!

"But this beacon also provides a direction for travel into the non-atomic worlds, into a dimension which carries man far forward in his evolutionary development into dimensions in which their worlds are completely and solely operative within a unified space/time continuum. Man, living in these higher frequency planets, is melded with the greater understanding of Cosmic Mind, which is the great transmitter for life on all dimensions. That beacon, which has been erected to provide for direction for truth seekers is known as Unarius! It contains the sum and total of the infinite for all travelers into the stars who have gained a foothold and have arrived at their destination and are proving their knowledge by giving of themselves to help the weary traveler over the rapids of his third dimensional beginning on the pathway to the stars!

"Many of these Brothers are scientists in the many related fields of earth knowledge. Many of these scientists are well-known personalities whose names have been raised and etched into the exterior walls of great buildings. Many of these humanists have been the architects and the designers of societies. These are the true scientists because they have indeed indicated by their own success and overcoming of the temporal world of time and space, the center post of self realization. Through the integration of the unified field of their mind transmitter, they have realized the equation of life that has so confounded third dimensional physicists. That equation in the exterior or third dimension is the physical anatomy which is one polarity of the infinite. It is the mind that is the other polarity.

"This, in essence, is therefore the real and full picture of life; for as soon as one can see through the smog of the material atomic elements that make up life, one will automatically tune to a higher frequency signal. That signal will be the transmitter of a higher, advanced being who once lived on earth. And that individual *searcher for truth* will, in this respect, have gained a greater ability to move ahead because he has now a beacon to guide him, more capable of determining the difference between the real and unreal, the mirror universe and the real universe. Thus such individuals will develop the necessary mathematics which will be his new mentality and he will be guided in this respect, ever so slowly but surely, until he gains the strength to form his own associations mentally with higher dimensional qualities.

"These are important developments that are resounding within the halls of science and are opening up new aspects and concepts of the reality that lives within the mirror of man's consciousness, because at all times, those advanced intelligent peoples who are truly the inspirational forces have been acknowledged by the man of science as his own, as the very teachers whose class they once sat in to hear of the widening conceptualization of evolution.

"There is no death. There is only life that is ever expanding and entailing new understandings, within a progressive evolution. This is the importance, and should be the present guidelines for the developmental nature of science in all disciplines.

"Let each and every scientist in his analysis, take into consideration at all times that he is involved in a fourth dimensional understanding whether it be in a sub-microscopic field or in the macroscopic fields of research in which he is functioning. Let him realize that as a *searcher after truth*, he is the subject and the object, and must realize he cannot objectify anything without taking into consideration his own understanding of himself as the subject in his research! Thus will he see the object of his research in its *true dimensional relationship!*"

Chapter 7

PARALLEL WORLDS

"The proposition of the two parallel universes is, in a certain way, not entirely a new idea. For many years a certain group of scientists have firmly believed and still believe that there is an exact opposite replica of our entire universe as it now presents itself to the physical eye. In other words, so opposite is this universe that it is the antithesis of our own universe, and should at any time these two universes meet in even a small way, they would completely disintegrate. Some proof of this theory is already in existence throughout the world and there have been certain very mysterious happenings, events or phenomena which cannot and have not been explained by any known scientific means.

"One of these events relates to a huge crater in Northern Siberia that happened in 1905. Some great object or thing came out of space and struck the Earth in that particular area making a crater some fifty miles in diameter and several miles deep. Oddly enough, there was very little if any earth or rocks or debris thrown up around the crater. Apparently most of the huge mass of displaced earth had literally vanished. This particular happening has puzzled science and, as of today, no plausible theory or explanation has come forth except that it was a piece of antimatter which came from the opposite universe and struck the Earth with such force and impact that the Earth literally vibrated. There were tidal waves and various seismic disturbances throughout its entire surfaces."

"Now the theory of the opposite universe is, to a certain extent, entirely true. As it was explained in previous discussions in this text, the dominant principle of creation in the manifestations of different interpolations is an oscillating recurrence between opposite polarities. In the third dimension this phase reversal is 180 degrees which manifests itself as an up and down motion called a sine wave . The fourth dimension equivalent is a 360 degree phase which unites itself in a complete and never-ending circle. However, within this circle there must be the same positive and negative polarities. In this respect then it can be visualized that this cycle is actually two cycles and that they are oscillating in unison with each other, reversing their polarities in the regular order of frequency oscillation with which they are incorporated. This universal pattern of cyclic oscillations is manifested throughout the interdimensional cosmos. It is the sum and total of this infinitely filled infinity.

"There are other very great and wonderful principles involved in this interdimensional cosmic hysteresis; that is, through certain frequency attunement, harmonic regenerations, etc., it can be visualized that vortexes, in a manner of speaking, form through the connection of different frequencies and harmonics in a chainlike fashion to form a kind of cosmic centrifuge of vortex.

"Now the constant oscillating phase reversals act in a certain way very similarly to centrifugal and centripetal forces. That is, due to the differences in the interdimensional frequencies, the higher rates of vibration represent a more positive bias to those of the lower. The lower vibrations then, in turn, can always be considered to be negative even though they contain their own phase reversals from positive to negative, etc., and so on down the line. As these harmonics and linkages continue throughout this vortexal pattern, the closer to the center, shall we say, of this centrifuge then the more negative do the various different oscillating phase reversals become in respect to those on the outer periphery or rim of this centrifuge.

"In a way, this is a kind of compression and it will ultimately result in a nucleus being formed wherein we have as sort of a sub-infinity, a high concentration of phase reversal energies which are purely negative in nature simply because they have approached and have begun to pass a certain line of demarcation which would, due to the frequencies involved, determine what is the fourth dimension and the third dimension. In other words, these frequencies are beginning to lose their cyclic patterns; then they begin to reexpress themselves in a different way by separating themselves from their cyclic patterns and become straight line frequencies of third dimensional frequencies which stem outward in a radial pattern from this central core.

"This is exactly the situation which is happening with our sun. Tremendous energy, through these oscillating processes, is being fed downward in a centripetal fashion into the central nucleus or core which is the sun which then, in turn, through the tremendous hysteresis or change within the wave form structures, as they separate themselves from cyclic patterns into straight line frequencies begin to radiate their energies into the third dimensional world or our solar system.

"However, it must be borne in mind that there are no clear, straight line points of demarcation between the third and the fourth dimensions. Rather, it is a gradual transition and can be envisioned as many different zones, in a manner of speaking, wherein certain things begin to happen in the natural oscillating processes. All of these differences are incurred due to the various frequency interpretations which

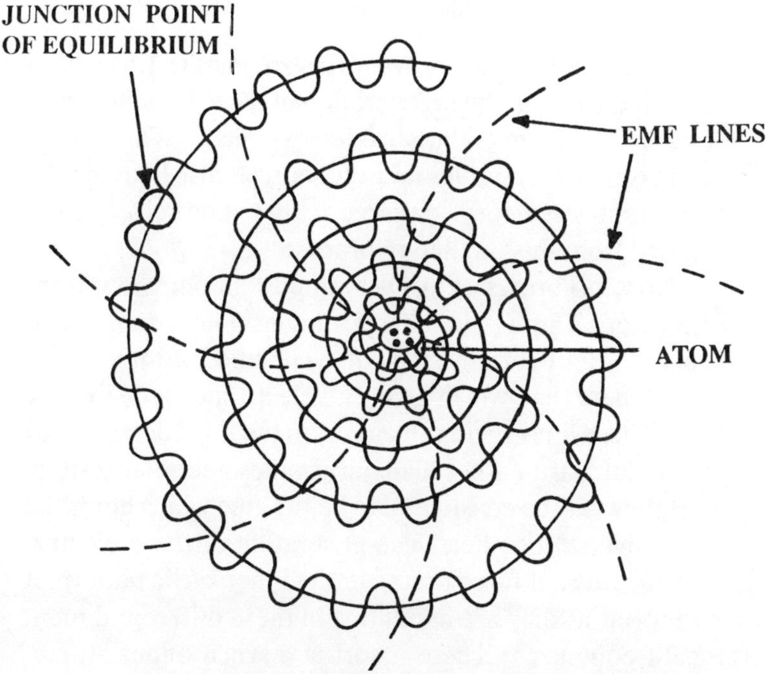

Showing EMF and wave form patterns of transconductance of energy from the 4th dimension. Atoms of steel (or the sun, or star, or galaxy) - all function the same.

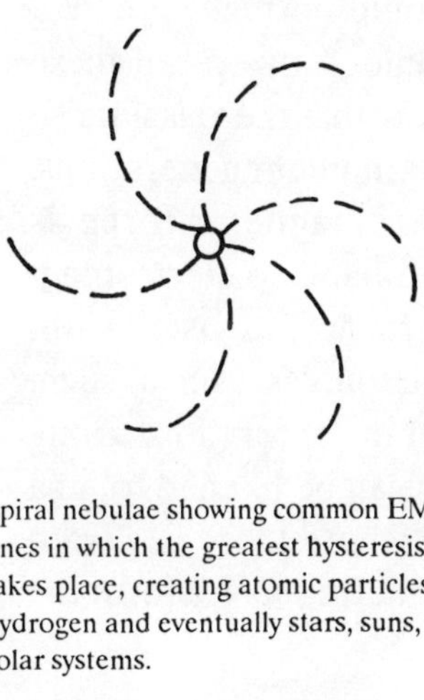

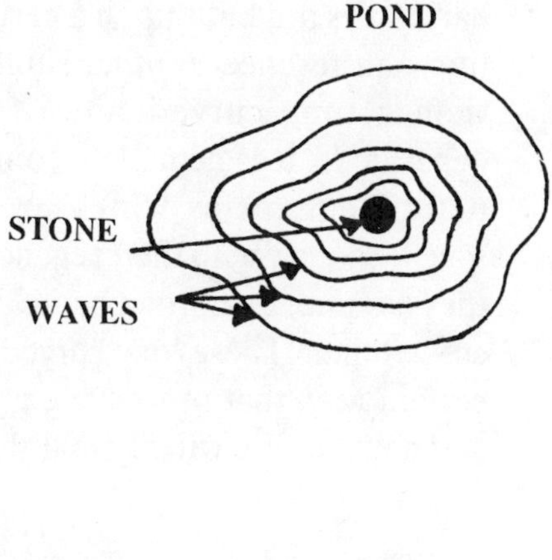

Spiral nebulae showing common EMF lines in which the greatest hysteresis takes place, creating atomic particles of hydrogen and eventually stars, suns, and solar systems.

these cyclic patterns and wave forms are manifesting within themselves; and in the regeneration of these harmonic patterns, again reform cyclic wave forms which, while bearing certain original facsimiles to there sources, also bear entirely different wave form patterns within themselves, compounded from this oscillation process.

"Now this brings us to a very important point in our introspection. There is, in the interdimensional cosmos a vast regeneration of certain kinds of cyclic lines of force, if we can call them that, which I will call electromagnetic lines of force. These lines again remanifest in the third dimensional around our Earth, other planetary bodies, the solar system; infact they may be considered to be the glue which holds the great universe together. The generation of these electromagnetic lines of force occurs in the inner cyclic pattern of wave forms as they are oscillating in these different dimensional frequencies. There is sort of a synchronous attunement set up wherein one given wave form in its cyclic pattern does immediately oscillate a particular given point in its net total circumference with a similar wave form.

"In a general consensus, this should normally create a secondary oscillation called a harmonic. However, if these harmonics are built up in a certain way that there is a pulse line transference to other similar harmonic regenerations, then a long curved line of electromagnetic force is generated. It is actually a composite band of electromagnetic frequencies which are operating an oscillation synchronously with their regenerating sources. And as such, they assume, as a broad band level of interpretation, a long curved line. These long curved lines can be likened to, in a certain way, that of weaving a cloth where there is a warp and a woof. We will say that the oscillating waves form the basic constituents of this interdimensional cosmos and form the warp. They regenerate cross fibers or the cross threads which we will call the woof. And it is these cross threads of the electromagnetic force which, in a certain way, helped to form and help to continually support the generative process in the net sum and total of the infinite cosmos.

"It is these electromagnetic lines of force which determine the pinwheel shape of our universe. These lines of electromagnetic force then, also hold the Earth and other planets in their respective orbits around the sun. The total spectrum of these lines of electromagnetic force also include such frequencies as are transposed into gravitational fields or gravity as well as in the electromagnetic spectrum, in the net sum and total of hysteresis as it is expressed by the conversion of solar energy into our heat and light spectrum as we know it here on Earth.

"This is all partially proven by many existing scientific factors; it is known that hundreds of miles above the pole temperatures are higher than they are the same distance above the equator in the area of the Van Allen radiation belts which are not really radiation belts but rather, they are more concentrated fields of lines of electromagnetic force which cross and crisscross above the surface of the Earth. It is in these electromagnetic lines of force - and there are many kinds of these regenerated elements of energy - which help to 'filter out' certain of the solar energies which could render life on the Earth nonexistent, such as the gamma or delta rays and which are very destructive to life on a planet. Now this filtration is not really filtration at all but rather, is a cancellation due to the phase reversals which take place within the wave forms themselves from the gamma rays and from the electromagnetic lines with which they come in contact.

"A note of interest that this socalled space about us is very densely filled with different kinds of electromagnetic lines of force; some of these are very broad and very powerful. Space ships or flying saucers use them as a source of energy traveling from one planetary system to another part of the galaxy. This can be done in a transposition of frequencies which involve the metals and other material which comprise the spacecraft. This craft can therefore be tuned, so to speak, to the movement of oscillation within the electromagnetic field and therefore travel at incredible speeds - perhaps several hundred times faster that the speed of light. And again in this hypothesis we begin to approach that inevitable astral flight proposition.

"Another way in which we can picture the total synchronous regeneration of lines of electromagnetic force would be something like a bucket brigade formed by firemen who lack pumping equipment and use buckets which they pass from one to another. In other words, the constant and multiple series of the regenerative process create tremendous lines of force. Some are, as might be expected, much stronger than others as they increase in the multiplicity of their regenerations and according to what particular band of cyclic frequencies which give rise and origin to them.

"This also brings us up to more elucidation on the principle and the concept of the atom itself as it compounds our terrestrial Earth planet. Now referring back a bit to our previous presentation which involves a tremendous interplay, a filling of space, so to speak, of a vast panoply of cyclic wave forms oscillating form huge centrifuges - that is, large in the extent that they encompass cyclic wave forms in many different dimensions or frequencies. Other cyclic wave forms may be very small and be harmonically attuned to the larger ones.

"As we progress down in space, so to speak, toward the ultimate formation of an atom, here we find a conglomeration of regenerative wave forms in cyclic patterns forming this same exact centrifuge but on a subminiature scale and in the net sum and total of the same oscillating processes, it forms the same nucleus. The nucleus, in turn, has, in its own respective dimension, all of the exact wave form counterparts which are contained in the centrifuge which created it. Then, as a means of expression, just as it did with the sun, a certain radial pattern of expression takes place into the third dimension. This creates a secondary electromagnetic field. Other lines of force which are exact replicas or facsimiles of the vortexal pattern revolve or regenerate very rapidly around this nucleus, but in the third dimension. As these numerous lines cross, they form parallaxes which also are synchronously tuned to the net sum and total and revolve in the same direction, as it concerns certain positive and negative phase reversals which are always inclined in one direction due to the difference in time between manifesting the positive and negative terminus of the cycle.

"Picture yourself pushing a wheel which is moving freely in space, and at a certain point, your hand slapping against the rim of the wheel would constantly accelerate this wheel in a certain direction. This is a kind of synchronous forward movement which causes the parallaxes of this atomic formation to revolve rapidly around the nucleus. It is also within these parallaxes that we find great concentrations of pure energy, oscillating again in sub-infinity in the exact composite or reproduced cyclic form of the original atom form. These charges of energy contained in these sub-infinite atoms are tremendously powerful and they can travel through practically any known so-called solid material.

"As an example, it is the charge of sub-atomic particles which, when used as a laser beam can drill a hole completely through a diamond in a fraction of a millionth part of a second. The reason again is very simple: As this tremendous charge of pure energy, revolving within itself and oscillating as tremendously as it does, meets carbon atoms which comprise the diamond, then there is an immediate cancellation of all wave forms involved and the energy reverts back into its interdimensional form. Matter can be said to have been disintegrated, as it is in the case where the diamond is drilled with a laser charge.

"In our introspection of the infinite variety and manner in which oscillating wave forms are compounded into the familiar atom-molecule combination, which go to make up our terrestrial Earth, we may be confounded and overwhelmed. Yet, all about us in this seeming, apparently solid, Earth there are countless billions of tiny atom forms which are regenerating harmoniously with their vortexal formations from the inner dimensions!

"If we add to this already overwhelming introspection, and with our mind probe into the inner dimensions, then this introspection is proportionately increased a hundred millionfold; for within this fourth dimensional consensus do we find infinity itself. Yet, this fourth dimensional flux field revolving as it does from its positive to negative polarities is a composite actually within itself of two universes; the terrestrial universes which we can see with our telescopes and our electronic devices, and the invisible and presently unknown universes which have created this universe as a manner and means of polarization in the interchange of oscillating energies between them.

"Then, if we extend this invisible infinity which has propagated and created this visible spectrum into more remote regions, we will not see at any point, any condition which would suggest matter and antimatter. While we might see such a vast difference that it would suggest one would immediately destroy the other, yet there is impounded within the creative law of oscillation as it takes place within the polarity phase reversals, that at no point and regardless of how far we extend our perspective into infinity, do we find an unbalance in any of these cyclic wave form patterns. Therefore, it is quite unlikely, in fact, impossible that this infinity, as infinitely intelligent as it is, in all its creative facsimiles, would ever arrive at a point where it would destroy itself.

"Rather, if there is any destruction which occurs either dimensionally or interdimensionally, then it is a proposition of individual choice as it might concern human beings; a choice which they make by becoming extremely self-centered and selfish by completely enclosing themselves in the citadel of their own ego. Plant and animal life on any planet constantly perpetuates itself in its common scale of evolution from seed to flower, etc. So likewise do all oscillating cyclic manifestations in infinity constantly regenerate themselves. In so doing, their cyclic manifestations perpetuate not only themselves, but they regenerate certain harmonic constituents which are constantly recreating new forms and perpetuating infinity to an even greater and even more expanded state.

"It is not possible in the human mind to conceive infinity; and should this be so, then infinity would cease. Likewise, should infinity cease, then all form and matter, either terrestrial or otherwise in the interdimensional cosmos would cease to exist, for infinity is the composite of all of this oscillating, regenerating universe of energy - if we can call infinity a universe and extend our perspective far beyond the limits of our comprehension.

"There is much more which could be discussed and most likely will be in further discussions, which will all tend to enlarge the perspective of your mind, an enlargement in the intellectual capacity, the facility to tune in to the different planes and dimensions of your life to gain access into sort of picture of creation as it exits around you and from within yourself.

"We have spoken of many other concepts which are adjunctive to the general perspective of infinity, - the parallel universe, or two universes oscillating within themselves are quite synonymous in many respects to you in your physical body and your psychic anatomy. For surely as you live consciously in this dimension with your conscious mind in a third dimensional world, so does your psychic anatomy live in a fourth dimensional world; and in a facsimile of oscillating wave forms in cyclic patterns, reproduce every facet, every tangible and many intangibles of your life. In this net sum and total process of oscillation between yourself in your conscious physical world and the world of which you know nothing - your psychic world - you are reproducing in this oscillating process the entire facsimile of your past and your present. Your future is also being reproduced which, when it is integrated with the time factor and separated in space, you will live through this future in the time ahead of you.

"In this way then, you will be contacting the infinitely numerous wave forms and harmonic patterns in their cyclic manifestations from the inner worlds. These will all be interpreted by your own conscious mind, in your daily life in that future as your life experiences. They will relay to you and into your psychic anatomy by oscillation, certain particular wave forms which, when combined with wave forms from your past, will again create the illusion of the picture world which you see in your mind - the picture world which makes your daily life possible here on the planet Earth. Should you grasp the meaning of what we are trying to convey to you in these words and should you be successful, then you will most surely live in a world which is entirely different from the world in which you now live. You will not live in a body composed of atom-molecule forms and supported by a metabolism which has been compounded from the beginnings from the protoplasmic forms which were first engendered upon this Earth by the cyclic movements from the inner dimensions.

"Just as it is now known to science of this world that certain molecules called amino acid molecules are the building blocks of larger molecules called proteid or protein molecules, the inorganic forms first spawned upon this Earth, like they are on many planets and as they understand it, came from bursts of energy as they call it. How much better it would be if they had used the word 'pulse'. A pulse synchronously tuned to a number of oscillating cyclic wave forms in the inner world and supported by certain harmonic structures, could group together certain atomic forms to form a molecule, a sort of synthetic molecule formed by the laws of creation, which remanifest constantly in a never-ending cyclic fashion on the basis of frequency from these inner worlds. As this first synthetic amino acid molecule was formed, then successive steps in this regenerative process would collect these molecules together to form a cell. This cell could be the first amoeba, tiny protozoa which, in its total life capacity, expressed hunger, expressed certain elements of self-protection - a tiny animal which could eat by literally surrounding its food and digesting it with certain enzyme substances formed within this tiny one-celled body. Yes, this tiny creature even had a circulation; it had certain parts which corresponded to a much more highly developed or evolved animal such as man himself - heart, lungs, liver, circulatory system - all first formed in the tiny vacuoles and other composite sections in this protozoa.

"No, the symphony of life, the entire harmony of life, its conception, its manifestations and its constant creation is not happenstance. It is not a product of spontaneous regeneration. Rather, it is a manifestation and product of infinite intelligence conceived millions of years in the sense time, beyond the periphery of its atomic, third dimensional form; an infinity which creates and makes possible all life forms and all manners and ways of living to those who can conceive this infinity, in its manner and way of breathing, living and remanifesting itself in its numerous planes or, as it was once called, 'The Many Mansions'.

"Curious as he may be and probe though he might, man will never go beyond the boundaries of this third dimensional world until he has conditioned himself from the inner worlds and from those great universities and teaching centers which exist in these inner worlds; the rigorous rhetoric of his third dimensional science limits the boundaries of his mental consciousness. And so it will be, the Earth will, to more or less and in a certain degree, always be a nominal plane of reference - a nominal plane of evolution - the formative phase or period of any earthman's life in his climb toward the infinite vistas of the inner kingdoms. And well that it should be this way. Were man, in the present time and in this generation, to possess knowledge and the ability to use this knowledge beyond the periphery of his intellect, he would destroy himself much more surely than he could by destroying the ecology of his planet.

"The ego must constantly be supported by the past. And at any given point, should anything materialize in consciousness which is beyond the immediate dimension of scanning, as it were, through the subconscious (the past), to make the necessary correlation between this past, then this immediate presentation must be destroyed or it must be shunned for man is a computer and is completely analogous to those machines which he has created and is now using in the industrial complexes of his world. So it will be in the future, until that time when man gains the mental ability within the consciousness of his own mind to tune in infinitely to one or any number of given points in this infinity, to manifest the cyclic motion wave forms into his consciousness as the sum and substance of his life.

"The material universe spreads across the vastness of space into an incomprehensible dimension, yet filled with pulsating life forms; countless millions of planets where countless millions of life forms exist. Some of these life forms would be incomprehensible to man on earth, yet all take their proper place in this vastness of infinity.

Chapter 8

DISCUSSIONS ON QUANTUM SOLAR MECHANICS I

"In our previous discussions we approached that vital and all-important subject relative to the interdimensional cosmos termed by Einstein as the fourth, fifth, sixth etc., dimensions, very important factors about the structure and transmission of energy. In general, scientists of today, cosmologists and astrophysicists recognize that space does contain plasma (sometimes called ether),a revival of a theory that existed some seventy-five years ago. However, now, just as it was then, no information or understanding of this space plasma is currently available. Limitations imposed by explorations in the vicinity of the interdimensional cosmos are most necessarily existent. The rhetorical minds of the third dimensional scientists, lack of proper instrumentation as well as other elements such as the requisites of preconditioning in higher worlds, etc., are most necessary before sufficient curiosity and desire is aroused which would stimulate explorations into this fourth dimensional cosmos.

"This space plasma was discussed to some extent in the two preceding chapters; however, the subject matter is most necessarily complex and deserves considerably more attention before some comprehension may be expected to be attained. We have found that space plasma, loosely defined by plasma physicists as a composite form of innumerable cycles of energy, all resolving endlessly around within themselves. We also discovered that these cycles, in the vernacular of the third dimension, could be considered to be small or large on the basis of their base plane frequencies on

Cosmic Rays or Particles of Charged Energy

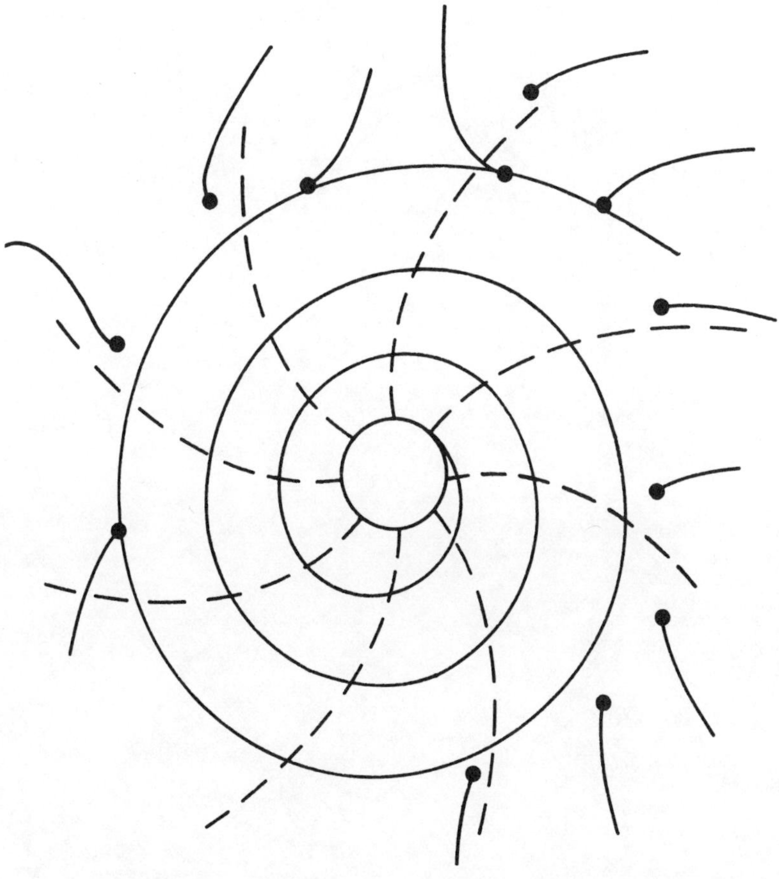

Showing how cosmic ray particles are influenced upon entering the earth's magnetic field, positive particles enter in an ordinary clockwise motion while negative particles are reversed and enter counter-clockwise, or according to the respective positive-negative bias charges of particles and earth magnetic fields. All such particles travel in curved lines following the lines of the vortex of the Universe. These particles are the compressed energy formations produced by the intense hysteresis between the sine wave vortex patterns and the electromagnetic force lines - exactly similar to the AC motor.

which they were oscillating. As each cycle is most necessarily complex, it does, to a certain extent and in a certain way, according to the frequency oscillations within the compound matrix, reflect the entire fourth dimensional cosmos. This it does on the basis of attunments, or similarities with different frequencies, harmonic patterns and a certain particular way in which the net sum and total of these frequencies and harmonic regenerations phase or peak, so to speak; a condition which can be called a synchronous isochronism.

"This generation of a complex synchronous beat of isochronisms is most important to understand because it is within this constant repetitious cyclic phase-beat that we find very important regenerations which, when harmonically attuned to other similar isochronisms generated in other similar wave forms, link up and form those most important structures that we have referred to as the electromagnetic lines of force. Considering space as we see it third dimensionally throughout our solar system and the galaxy, there is a considerable amount of inductive transmission going on at all times. This gives rise to the appearance of hydrogen atoms. The beat frequency oscillation refers to, of course, particular fourth dimensional densities which, when also linked up, give the centrifuge pattern resulting in a certain compression into the nuclear form which can be a hydrogen atom. The proposition of beat frequencies, isochronisms and the regenerative harmonic structures in cyclic patterns throughout the cosmic universe, is also responsible for transmissions of energy from solar bodies as suns and stars which have been cataloged throughout our galaxy.

"All stars and suns have this particular similarity. They are the nucleus of an atomic centrifuge. Some exist in the third dimensional sense and are much larger than our own sun; or that they are comparatively newer or that they are older. They radiate frequencies differently, however, according to how the net sum and total of the interdimensional centrifuge is compressing energy into them.

"Now one of the facilities of this interdimensional centrifuge as it concerns the nucleus, is the in-between fourth and third dimensional stage where there occurs this regeneration of isochronisms in cyclic forms which gives rise to numerous lines of electromagnetic force. If these lines could be visualized with the eye, you would see the sun radiating a tremendous field of energy in a curved or radial pattern, just similar in configuration to the familiar pinwheel formation. Lines of force as they are determined in their radiations from the sun follow these same curved patterns. This is so because the sun, radiating energies into our third dimension, but can be considered to be only one polarity; as there is in existence within the cosmic centrifuge, and at a great distance we might say, in terms of frequency, a second sun - an unseen sun - which forms the positive polarity. Similar in formation to the common and familiar bar magnet, these electromagnetic lines are stemming out radially in a pinwheel shaped pattern from our positive invisible sun to the negative, visible polarity-sun.

"The same situation is quite similar with all of the planets within our solar system. They also have a secondary counterpart or polarity within the fourth dimensional cosmos to which they radiate these radial lines of force. Again the same familiar pattern is formed, which we can see by placing a piece of paper on a bar magnet and a pinch of iron filings on the paper; when the paper is tapped the particles of iron assume that radial pattern that is so familiar to us.

"Now we can also carry this same concept into every one of the star bodies within our galaxy or our universe, and they also have a similar polarity - call it negative or positive, whichever you prefer - to which they oscillate interdimensionally. And we carry this same facsimile of concept on down to the hydrogen atom. From every visible atom as it is compounded into molecules and our familiar everyday substances, there also beats with the atom in this structure, an interdimensional atom which can be considered to be its other polarity. This is the true concept and meaning of the parallel universe. We find these same structural conditions existing throughout infinity to whatever plane we wish to turn our introspection.

As was once stated, we must always have two polarities within any oscillating condition wherein we have a phase reversal. This is a universal concept of creation and carries the idea or information 'consciousness' of wave forms, to one extreme and then to another extreme - a condition of 360 degrees phase reversal. In our third dimension, this phase reversal is separated by 180 degrees and results in our familiar sine wave equation which powers all motors and makes possible the lighting of our homes, turning the wheels of factories, and powering the most important and necessary functions of our jet aircraft.

"To whatever and wherever we turn our attention, we will see the familiar sine wave at work. Sometimes the sine wave is purposely degenerated into one-half a phase or a 90 degree phase which is called direct current simply because it does not manifest a second polarity. However, any metallic substance such as a wire, will present great and tremendous resistance to this direct current wave form. This is a common condition in the transmission of electrical energy over a wire as we do in interdimensional space; that is, the molecules and atoms which comprise the copper or other metallic conductor must conduct energy in exactly the same way that the many oscillating cycles which comprise this plasma are operating in that interdimensional space between us and the sun. This principle is exactly the same.

"In its energizing capacities, the sun is radiating its energy into the third dimension, and as was stated before, there are certain zones wherein differences in wave form structures are manifested; and are oscillating in many different ways simultaneously. Some of these transmissions most vitally concerned are the Earth worlds and in the electromagnetic fields which the sun has energized. As the Earth represents a certain polarity in this cosmic hysteresis, it is then coupled in this frequency interplay with the sun, as the sun being a stronger polarity or stronger source of energy, does to energize such secondary polarities which are the different planets revolving around it.

"Here again we find another very important function always taking place; how the orbits of the planets themselves are rigidly held in these orbits, but not necessarily from any physical laws as we see in activation when a satellite circles the Earth. In these instances, the centrifugal force generated by the speed of the satellite exactly matches that of the gravitational pull so that the satellite can circle for a long period of time in what is called free space, without meeting any of the restrictive factors which would ordinarily slow it down, such as friction with the air, etc. However, with planets that circle the sun, it is the electromagnetic lines which hold these planets into their respective oscillating positions, electromagnetic lines of their own particular polarities which are concerned with the fourth dimension as

Striking view of the solar corona prepared from data supplied by NASA's solar Maximum Mission Satellite. The colors represent densities of the corona. The spike extending from the densest coronal region (purple) persists beyond a million miles from the surface, and can extend to one-tenth the distance to the Earth.

well as the polarity - positive and negative - of the sun itself. The whole interdimensional oscillating structure is extremely complex and it will take a considerable amount of constant introspection before a sufficient amount of comprehension can be obtained as to this oscillating condition.

"Now you no doubt have heard and seen, especially in the eclipses, some information on the chromosphere of the sun. The chromosphere is that particular envelope which radiates, or at least apparently seems to radiate, tremendous energies and contain solar prominences that can be seen to be erupting a hundred thousand miles or so into space. This tremendous energy radiating so freely as it does is called atomic fission or that it is being burned in an atomic furnace. Of course this situation is not quite true. The scientist of today does not know that the sun has energy 'pouring' into it from the interdimensional cosmos. He also does not know the true factors concerning the conversion of energy into respective atoms such as the hydrogen atom which is said to be a fuel burned by the sun, and its conversion into a double configuration in its atomic form of helium. Also noted is, in a spectroscopic examination, that the sun contains 66 of the elements of the Earth. This would of course be a contradiction. We know that such elements as uranium, lead, radium, etc., aside from being to a certain extent radioactive, radium in its completely degenerated state, becomes lead. Uranium can also be made into such isotopes as plutonium, U-235 Americanism and other artificial isotopes.

"A certain rare element hardly ever found on Earth has been found on one of the great stars in our nearby galaxy Andromeda. This rare element promethium, and the way in which it was found on the surface of this sun by means of a spectroscopic examination, entirely defeats former concepts held by astronomers and astrophysicists as to the nature of the elements which are found in the sun and in their respective positions. Finding it on the surface of this star in Andromeda has certain very important connotations. Formerly it was believed that all chemical changes occurred deep within the structure of the sun or the star; however, the finding of promethium on the surface of this star invalidates this concept.

"In other words there is no other explanation, other than the fact that suns are constantly changing their chemical composition as they age. Here again is the assertion of certain facts relative to cyclic transmissions. Age, while it is a third dimensional factor relative to the transmission of time, again becomes one of those cyclic transmissions or *phase-beats* which we find in the interdimensional cosmos. In other words, the sun or star as a nucleus begins and ends its total life span as a simple beat frequency, wherein there is a total manifestation, we shall say. The total regeneration of the sun from its beginning to its end is the total consummation of this phase frequency or beat frequency which was generated in that cosmic centrifuge.

"Similarity, all such elements that have been discovered in the sun, are not necessarily found in any particular layers or within certain portions of the sun. Structurally speaking, if we could slice a sun in half, rather than layers, densities of frequencies, the way in which such frequencies beat harmonically in the regeneration of cyclic forms, etc., we would find such solid masses which might be relative to rocks which form mountains, etc., in our familiar earthworld planetary configurations. These are very important facts about the interdimensional cosmos, the way in which tremendous energy is regenerated into many different planes or into many other different dimensions.

"We can unite not only galaxies and universes but also discover how universes themselves become only mere microscopic specks in a more total infinity. And could we find words and could there be sufficient comprehension of these words, there is no doubt that such explorations could be continued indefinitely. This is indeed, the proposition of infinity, its consistency and its mechanics.

Chapter 9

DISCUSSIONS ON QUANTUM SOLAR MECHANICS II

"Our preceding discussions related to the interdimensional cosmos, the various interpolations, mechanics, etc., which are involved in the net sum and total of all processes which can be called infinity and also which produce our third dimensional world as it is viewed on the planet Earth, the solar system, the galaxy, the universe, etc. Some of these presentations and the concepts therein may seem, especially to the scientists of this world, somewhat farfetched or in the vernacular, 'way out', otherwise unsupported, etc. So to serve all purposes best and to use the ever-abundant proof that these concepts and presentations are indeed true and accurate, we will use the very essence and nature of the earth world science as it is presently known, to validate these presentations.

"As a beginning in our central objectivity we will use that closest heavenly body to our earth, the sun. Aside from the earth itself, there has been, no doubt, more study and exploration of the sun than any other body which presents itself in our third dimensional cosmos. Using the Encyclopedia Britannica and quoting from it, as a point of reference, we will enter into a discussion which will prove and re-prove all presentations and the quantum mechanics contained therein, in all presentations which we have so far made. As we know, the sun is a tremendously energetic radiating body of energy; the exact amount of energy which the sun radiates every hour of the twenty-four hour cycle or any particular given hour as it concerns the earth is indeed enormous.

"In a single day, the sun radiates enough energy on the Sahara Desert to power all of the earth's needs. Now this has always posed a very important question, not only to primitive man but to mankind in whatever particular civilization or status quo in which he has existed. To our present-day scientist, it is even a more enigmatic parable than it was to the primitive man who could more or less deify it, call it the Sun God, etc. The scientist of this time, however, is much more concerned in the parable of what the sun is and the source of its enormous radiations.

"The context of information contained in the encyclopedia contains many contradictions, either apparently not explained or neglected, or otherwise, that there were no suitable explanations to eliminate these contradictions. For example, the source of energy from the sun is supposed to be contained from the net sum and total of radiated energy from the implosion of two hydrogen atoms. Now an implosion is exactly opposite to an explosion; that is, the two atoms would somehow have to collapse together and form a totally different atom, which is called helium. In this total collapse there is a large amount of energy released. However, in this theory the scientist has not yet explained why there are more than sixty-six other different atom forms contained in the photosphere of the sun; iron and nickel being among these atom forms as they are cataloged in our atomic weight here on earth. Now if the net sum and total of the energy released from the sun is used what are the other atoms doing in that photosphere of the sun?

"The scientist also does not rationalize or give a suitable explanation of how this implosion process is conducted, what instigates it, and why it is that a helium atom is twice as heavy on the atomic scale than were the two hydrogen atoms, despite the release of considerable energy; in-fact, energy which reasonably could be supposed to be at least equal to the net sum and total of their atomic weight. Also quite evident, how is this process conducted and how is it governed?

"Here on earth the scientist can produce the twenty-five million degree temperature which he believes is necessary to conduct such an implosion process; however, he has not yet constructed a magnetic bottle to contain the atoms in this process. Of course the scientist has the entire process backwards. The twenty-five million degree temperature is actually the quantitative release of energy from this implosion. It is not necessary for these two atom forms to have a twenty-five million degree temperature in order for them to merge. Rather, it is that they should have some other intercosmic law or force governing their actions which would cause them to merge in this metamorphosis. And in this merging there would, of course, be that quantitative measurement of energy released. The twenty-five million degree temperature is merely a theorem; it has not actually been confirmed.

"The scientist is like-wise mixed up on what is considered heat. There are ninety-three million miles of space between here and the sun. There is no medium in this space which would transport heat or conduct heat from the sun to the earth. Therefore, in the absolute consensus of energy releasements as they were concerned with two hydrogen atoms, any conductivity through this third dimensional space between the sun and the earth must be done in an entirely different way. Therefore, this energy is not heat, and the twenty-five million degree (supposed) temperature on the surface of the photosphere of the sun is really nonex-

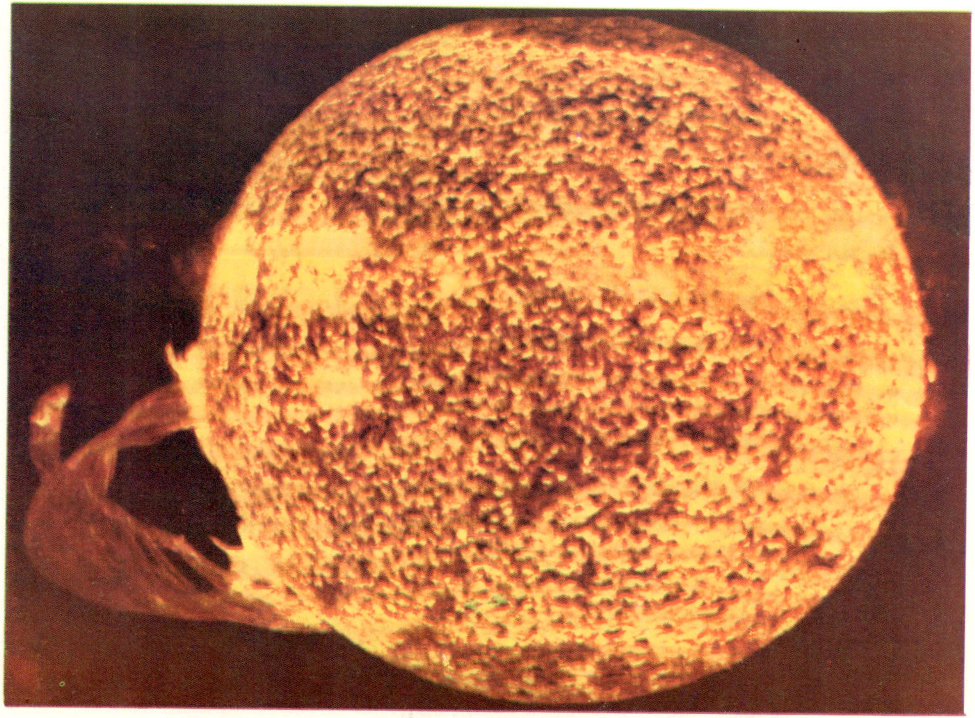

One of the most spectacular solar flares spanning more than 367,00 miles across the solar surface. The solar poles are distinguished by a relative absence of super-granulation network and a much darker tone than the central portion of the disk. U.S. Naval Research Laboratory, NASA's Skylab 4.

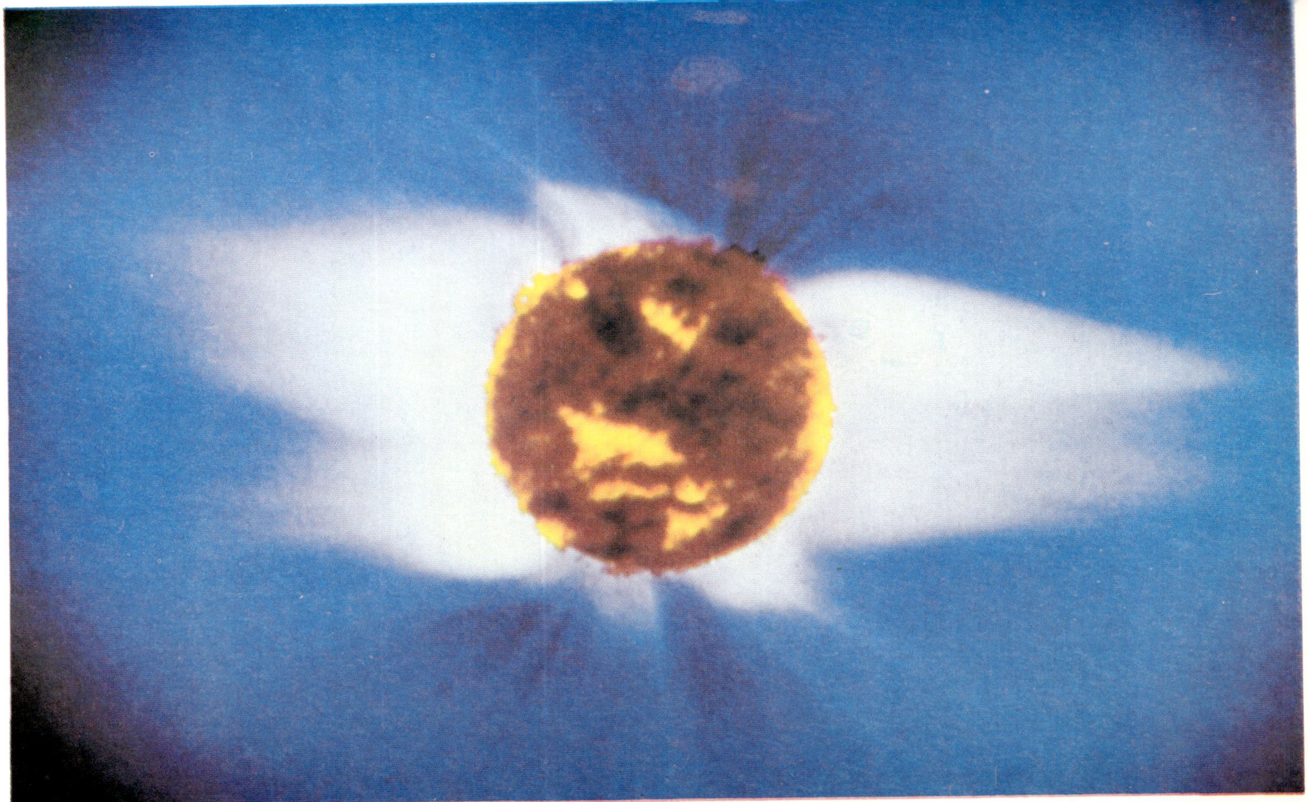

Solar X-ray Rocket Experiment 170 Angstrons. White Light Coronal Camera - 01:03 GMT, 18 March 1988 High Altitude Observatory/National Center for Atmospheric Research. University of Colorado/Laboratory for Atmospheric Space Physics istent; it merely resides in the mind of the scientist or on his papers as a figurative consensus of energy which is being released and is not really heat, for in a pure and abstract sense, heat is nonexistent. It is merely the measurement of certain energy wave forms which, due to their specific size, can be quantitatively assayed to have a certain resistive factor when they pass through certain solid particles which comprise our third dimensional atomic constituents.

"In other words, an energy wave form passing through any given mass meets with a certain resistance; and this resistance is the sum and total of another form of energy which is released, which the scientist calls heat. If the wave form passes through our skin and our flesh, and providing the wave form called heat is of sufficient intensity, then we shall be burned because the cells in our tissue structure must absorb this "heat", because they are resistive to it. In absorbing this heat the cells will explode and be destroyed, therefore, you would be burned. Smaller amounts of these heat wave forms can pass into the tissues and create a sensation of comfort as the body is normally maintained at about 98 degrees Fahrenheit.

"So you see, the scientist of today is still in a rather primitive state of evaluation as to the net sum and total of his consensus in the different ways, manners and means as well as the energy wave forms themselves in determining what should be most properly called what.

"The photosphere of the sun then, becomes a radiant source of solar energy - spectral energy which, to a large degree, would be almost totally unknown to the five senses of the earth man. It is through the various inductive and conductive processes, called cosmic hysteresis, that we find this energy converted into the more familiar forms with which we are so familiar, here on earth, including heat, light, and the electromagnetic spectrum which includes gravity as well as the magnetic spectrum itself.

"One more proof is the magnetic spectrum itself. Now the sun is about 334,000 times larger than the earth, yet measured by the instruments available to the scientist, it has a much weaker magnetic field than does the earth, based of course, upon the proportional size of the sun and that of the earth. If all equations were true the sun should have a magenetic field 334,000 times stronger than that of the earth because it was that much bigger. What is really implied, is that the scientist measures his magnetic earth field or the magnetic field of any solar or stellar body on the basis of a rather crude implementation. He measures only one basic part of that magnetic spectrum. It is more or less a physical property more closely related to the total sum of atomic forms as they are contained on earth, or from the total spectral electromagnetic field. There is a similarity here with the encephalograph, in measuring brain waves; a scientists, although he may think he is measuring brain waves, he is however measuring only a subharmonic.

"In other words, the scientist of today in measuring the magnetic field of the earth is measuring only one small bit of this electromagnetic field - a sub harmonic, if you will. Because the sun is much more totally involved as the radiating source and because of the tremendously greater power, and as the central centrifuge in the solar system, the electromagnetic field is still in the higher reaches, shall we say, of electro-radiant forces which are part of that fourth dimensional centrifuge and cannot be measured by atomic instrumentation!

"Now part of the radiant process of conversion of energy from the sun, the scientist believes, is that about one percent of the total hydrogen which comprises the mass of the sun would furnish sufficient radiation from this implosion process to power it for about one billion years. Now that is a long time. But of course what the scientist does not say is where did the first hydrogen atom come from and again what caused two hydrogen atoms to merge. Also contained in his hypothesis is the ping-pong effect of different atoms which comprise the photosphere of the sun. He calls these various elements, as he has cataloged them such as iron, nickel, calcium, potassium, etc., which he has found in his spectroscopic analysis as actually being elements in that sun's photosphere. He says that these elements are releasing electrons because they are ionized. As they release electrons, other atoms around this atom pickup these electrons; and in this releasement and 'pickup' process there is a releasement of energy. Also, that the atom which has picked up the stray electron has reciprocated and has batted his own electron back to the first one.

"Now if this was even partially true, we again could ask this very important question: Why do atoms do this and what causes them to reciprocate? Well, if the scientist understood what we have presented in our previous discussions, he would understand that this is actually an interchange on the basis of synchronous isochronisms between two atom centrifuges of energy, which are releasing energy into the third dimension as a result of this synchronous interchange of wave forms between them. The laws which govern this interchange have been well-defined. We have called them the regenerative principles of harmonics, or harmonic regeneration. In properly tuned sequences, which we can call synchronous regenerations, we will regenerate a net sum a total of a different set or band of frequency regenerations called isochronisms, which beat either regularly or irregularly, according to their basic and formative constituents.

"Here again we cannot overemphasize or stress firmly enough the importance of the interdimensional cosmic principles which govern and control all of these interchanges and formations. One other factor concerning the sun which puzzles the scientist is sunspots! Now through proper filtration, he can view the surface of the sun to a degree where he sees that its surface resembles a leafy pattern; that is, like a pattern made up of small leaves glued all over the surface of the sun. This has puzzled him. He has also found, in the regular cyclic period of twenty-two years exactly equated as a one-half cycle of eleven years, the production of sunspots. Whereas, in the recessional part of that cycle, the sunspots recede and diminish so there are very few if any of them left. The scientist has well marked the appearance and reappearance of these sunspot cycles and it might be a note of interest to interject at this moment, that from out of the Atlantean and Lemurian epochs there emerged this same cyclic transference or that the same interdimensional science was taught at those two particular times as it is now being taught. The eleven and twenty-two year sunspot period became a great mystical symbol in all of the different arts and sciences which existed from the Atlantean time, in their own particular forms, degenerated from out of the original Lemurian concepts.

"The central corridor in the big pyramid in Egypt is said to be called the corridor of time, and prophetically marked every important event in the earth's history from the time of its construction down to the present time. It was based on the sunspot cycle. That is, the Ancients knew and used the determinant in their astrological and astronomical sciences as they concerned the interdimensional cosmos, to prophesy what events could occur on the earth as they concerned mankind. They became very true indeed, for it is in these different cyclic patterns in the interdimensional cosmos, that we find are of the greatest influence, not only on this planet earth but in any other particular planet in whatever galaxy we may find!

"Life is always indigenous and productive according to the net sum and total of energies which are propagated or reflected into the third dimensional world through the processes which we are describing. The psychic anatomy itself, of every human is with every living thing on the surface of the earth, as subjective to these intercosmic principles, and a human will flourish according to how these principles affect him in his daily life. This is the true meaning and concept of astrology, not that any planets themselves do or can influence us to any appreciable degree in our earth life, but rather, it is in the interdimensional principles which govern the interplay of this intercosmic infinity, that are governing elements. They control not only the planets but they control everything which they have produced as part of the net sum and total of the productive infinity which is regenerating infinitely; and produced all known life forms and all unknown life forms, and will progressively re-create unknown life forms even into infinity.

"When this proper understanding of intercosmic laws concerning infinity and the interdimensional cosmos is more fully understood, then our society, the governments of the world could be proportionately and more intelligently constituted, and be used advantageously to the betterment of all peoples concerned.

"The astrology of today is primitive and is based on some rather degenerated accumulations which were deprived from the Chinese astrology at that time, contained thirteen different signs, all based on the appearances of different configurations in the heavens. Now anyone who has a telescope or who has read a book on astronomy can tell you that any of these configurations - as they resolve into astrological figures such as Sagittarius, Aries, Pisces, Scorpio, etc., represent the configurations as the Ancients found them in the stars of long ago. Any of these configurations are actually composed of many thousands more of unseen stars of different magnitude which cannot be seen with the unaided eye.

"Now since the time of the original astrological concepts as they are contained in our modern astrology, there have been many other different discoveries in our universe which have been made by astronomers; all of which have tended to invalidate any particular astrology now currently in use, on the basis that any and all other heavenly bodies, otherwise designated in astrological charts, could be just as influential as were those which had been so designated. Also, in several thousands of years, star constellations have moved or we have altered out position to them, invalidating the astrology concept.

"We can extend this hypothesis much further if we interject our own particular interdimensional cosmos and, as we have done in preceding chapters. The vast, infinitely filled infinity, a plasma or an ether, as it was formerly called, would resolve itself into an extremely complex pattern of cyclic motions of energy, each one infinitely complex in itself and revolving endlessly within itself, repeating its own information - all harmonically linked to each other, regenerating

An Atom or A Universe

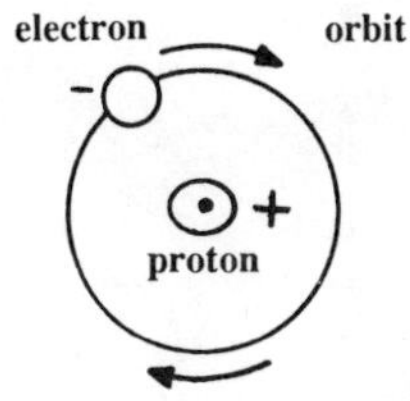

A Simple Hydrogen Atom

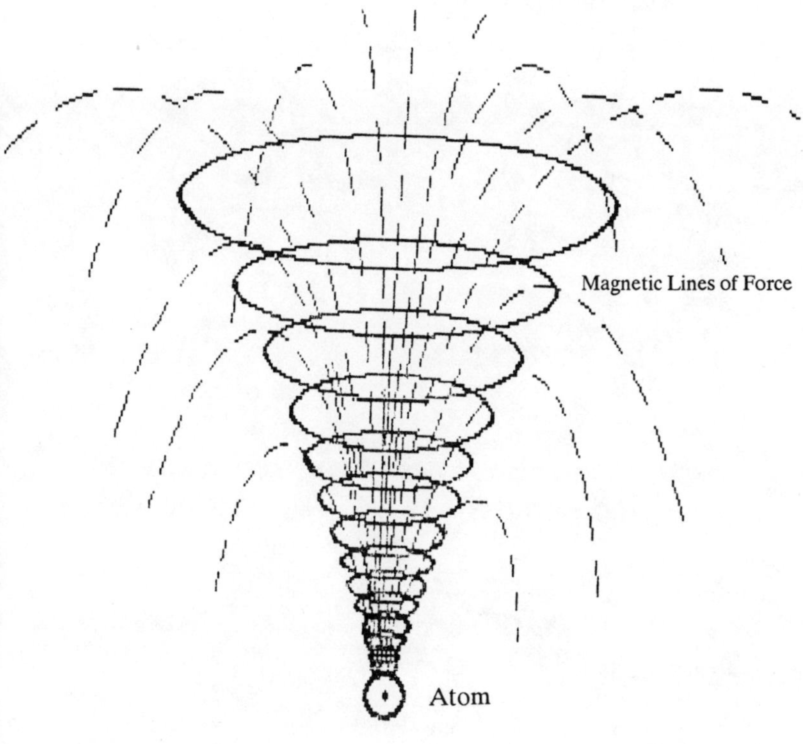

synchronously certain different wave lines of force, to form atomic constituents.

"Now getting back to the atom! The astrophysicist of today knows that his so-called space contains billions of tons of finely divided atom particles of hydrogen. He has theorized that it is in some centrifugal motion, that has gathered together a large quantity of these hydrogen atoms and formed a sun. Now if this is true, we should ask this very important question. By what manner and means and who commanded these atoms to gather together in a centrifuge to create that sun? In a sense though, the scientist is correct in saying that it is a centrifuge that creates a sun, whether it is our own little dwarf sun or it is one of those huge, enormous suns such as are found in Andromeda or in the Constellation of Orion. And by what manner or means were these singular hydrogen atoms formed? Why are they found in space? Well, the answers there, too, are found in the quantum mechanics of the interdimensional cosmos.

"Each atom represents a miniature centrifuge of energy. In this central energy centrifuge, the hydrogen atom becomes merely the third dimensional nucleus or core. On the other end of that oscillating centrifuge is a similar cosmic atom which is not third dimensional in nature but which in an oscillating process, regenerates and passes back and forth, so to speak, in its oscillating process, the net sum and total of what is being regenerated by all of the wave forms contained therein which resolve themselves in our present-day atomic scale of weights as the hydrogen atom.

"By whatever manner or means we analyze our present-day sciences, and the present-day perspective of these sciences, we find enigmas and dilemmas in the different scientific presentations. By the mere extension of our perspective into the interdimensional cosmos, we begin to recognize and to understand why these scientific aspects exist in their enigmatical form as they do with the scientists of today. Whether the rhetorics of this world prohibit the mentality of the scientists from escaping this third dimensional world, or whether it is pure stubbornness, or by whatever other name you wish to call it - a lack of imagination, preconditioning or whatever - it is quite true and apparent that the scientist of today is even in a much more abysmal state of ignorance than any of his predecessors; even stemming back into the more aboriginal state beginnings of existence. And this can be substantiated on the basis the each new scientific discovery which he makes completely invalidates certain previous discoveries he has made. It also presents the inevitable enigma that he does not have sufficient intellectual capacity to escape into the interdimensional cosmos where he could fully and understandably equate not only his third dimensional world but also the entire concourse of his evolution.

"There is one more final bit of evidence which we wish to present which, in itself, is one more validation of these interdimensional concepts. This is in the structural entity of a sunspots itself. Through his telescope and by various filtration means, the scientist knows a sunspot is a centrifuge, very much like a cyclone is on earth. There is energy radiating from the center of this sunspot toward the outside edge or rim in a round whirlpool-like fashion. In a similar and likewise manner, energy from the upper surface of the sunspot is whirling around and around and descending into the center of this sunspot. The sunspot itself looks dark in comparison to the photosphere which surrounds it, however it is very bright indeed; and so far as can be determined by the scientist, it is energy just as is the photosphere. Now whatever constituents it is composed of are still comparatively unknown.

"We can illuminate to some extent to the scientist of this time, a fourth dimensional consensus, wherein we find the centrifugal patterns of frequencies, interplaying as they do, down into the sub-infinity to form a similar polarity in the opposite direction. Again to quote Einstein; 'if a line, is drawn far enough, it would meet itself on the other side'. In this concept of sub-infinity, we can again equate the proposition of a parallel universe. Or, we can again resubdivide the parallel universe into four. We can again square these different centrifugal patterns of super infinity or sub-infinity into infinite proportions which will give us somewhat of a perspective of what infinity really is!

"Here again, by whatever manner or method we use to extend the dimension of our mind into infinity - whether it is into sub-infinity or into super-infinity or by what other term you wish to subdivide infinity - means that in a general abstraction there are no limitations as to how far or whatever direction we wish to go. We will find in whatever direction we progress, in the structural sense, that infinity as we so find it is constantly deploying itself according to the law, order and harmonics which we have described in these previous chapters. And in the manifestation of these numerous and infinite regenerations do we find the structural entities of universes, subdivided into galaxies and again resubdivided into planetary systems. In this subdivision, planets are themselves being subdivided into atomic constituents going on down into sub-infinity and reexpanding in sub-infinity.

"This information of the plasma universe is far beyond the dimension of the third dimensional mind, however, it is a challenge- a challenge which must to some degree be met and equalized in the consensus of our own intellectual processes. And if we wish and desire to perpetuate our own life in a logical and an intelligent life cycle, then it indeed behooves us to use all logical manners and means, all information, all ways to learn of this infinity. And conversely, as we learn of infinity do we perpetuate our immortality.

Chapter 10

FROM QUARKS TO QUASARS

"The age of fourth dimensional logic and reason is fast approaching. But for certain reasons based on the surcharge and the discharge of previous unmapped territory of the mind, there was an unwillingness to confront those social forces that had developed beyond and above the concern of man's developmental nature, to add to the consensus of interdimensional knowledge of man's place in the universe. Therefore, innovations in mathematical theory and linguistics, in astrophysics, cosmology and philosophy, have become a minor factor within the larger established framework of attitudes that make up the present social systems of planet Earth.

"The various groups of individuals who have been either supposedly elected by their peers, to represent them because of their abilities to see further into the nature of life, have not always been a representative of the minority group, those who were the advanced architects, bringing new value systems which would uncover the pollution that had been sunk deep into the psychic nature of human nature which had blocked and restrained him from such knowledge. These futurists were not able to add, in a full and positive developmental and incremental manner, the awareness of mankind as a particle in the larger protonic forces of nature, within which each individual is so evolving as an electronic force. The protonic force, the central nucleus which represents the creative forces, and is the causal factor for all electronic forces that are the visible reflection of third dimensional life, have not made an addition to societal life to the degree that it reflected through the atomic architecture of third dimensional planetary life.

"At the present time, various disciplines of the scientific community are cooperating to a greater extent to gain an understanding of the unified nature of the forces of energy. Astrophysicists, cosmologists, and particle physicists are sharing information of their respective fields, which are not separate but are part of the whole. The whole, however, has been the mystery, because it had been separated and studied individually instead of recognizing that the electromagnetic lines of force bind all factors of the infinite vortex! This could be drawn as an example from the sociological implications of interrelationships between people, in work, in family situations and in the triangulation of mankind.

"These electromagnetic lines of force are always functioning and are being recognized from the gross perspective to the more subtle. The more subtle perspective or reality is within the research now being carried on by particle physicists and cosmologists who have been motivated to join the four forces of the energy of the atom under "*The Grand Unified Theory*". A grand unified theory would explain these forces and their symbiosis by giving an explicit, and definitive understanding with implications of the cosmological theories of the creation of the universe.

"Cosmology contains the fuller implication of an inflationary universe which would ratify the first quantum moment of the universe, the Planck moment 10^{-43}. The groupings of star systems in galaxies and the clustering of galaxies provide a larger perspective explaining the existence of virtual particles and indicate the serial development of evolution.

"In other words, science is attempting to find out about

the causality of nature by moving backwards in time towards that period or cycle when all factors that contain the features of the present universe had been composed. However, since all features of the present universe are derived from the fourth dimensional vortical field, there will be a dichotomy in present high energy physics research in its intention to understand the initial nature of the creation of present third dimensional protoplasmic energy.

"As has been evidenced and indicated in previous chapters, the universe is a composition of a hundred billion galaxies which are moving at incomprehensible speeds in their expansion increment.! Should any mathematician attempt to resolve the incremental nature of this expansion in some form of mathematical equation, they could not arrive at the initial wholeness of the number because the sum of the parts do not necessarily represent the whole. In fact this wholeness takes on new consequences and does not relate to the mathematical indices that were the basis for the calculations to attempt to put back into one whole equation the separate integers to explain the meaning of a mirror or parallel universe.

"Because of the physics of energy, once the apple has fallen and has been broken into various incremental parts, it cannot be returned to its original form by gathering up the pieces and returning them to that point from which it fell, due to the forces of gravity within the electromagnetic spectrum. In the same sense, all factors of the third dimension, being proposed by those now actively involved in strategies to combine the individual pieces to understand the whole, cannot be resolved because the whole is not only larger than the parts, but is a different number, since it is now occupying a relationship larger than the inverse ratio of its protonic parts.

"The fourth dimension is a realm, a region, a world, if you wish, a state of mind which is not only the sum of all the parts which is considered to be an energy index, but no longer occupies a field that can be equated by the measurement of its mass because it is not of an atomic design. In other words, the measurement devices of astrophysicists, cosmologists, and particle physicists, would fail to detect on their instruments any noticeable signal which would indicate the presence of energy (i.e., virtual particles). The reason for this is that present instrumentation is the projected development of scientific knowledge equated on the basic calculus in which third dimensional life has taken on its recognizable features.

"But as man himself contains the answers to the puzzle and, in many respects, is the puzzle within the puzzle of life, it is only through the development of each individual's awareness of his fourth dimensional nature that he can develop the system, if it can be called such, to probe into an area that is not based upon third dimensional measurement devices. Naturally, the entire purpose of life would be stymied if there was no possibility of learning of the fourth dimension, at least to the extent that there is a possibility that it exists.

"It is the existence of this field force that beckons to man, which is his libido and drive for life, and is the basis of serious research into the nature of matter! Naturally, the means of research must be with different implementations which presently are not accepted in the scientific community. In many respects, this is all well and good, because the slow and steady development of the *nature of reality,* must be ratified on a scientific basis and accepted by those individuals who, as professionals, are working within a system that functions under the auspices of logic and reason. That is to say that a hypothesis once advanced, before it can be validated as a reality, must be proven, not only once or twice but a hundred or more times. It is in this means that the establishment of science advances by collecting evidence and through experiments, validating the theory advanced as to the nature of reality.

"Evidence that has finally been resolved into a new understanding of man is now considered to be fact. It is a concept which, because it is workable, can advance the life of society. Society can advance only if individuals within the social structure, themselves advance, because the society is the proof of the changes that are taking place and of the direction of each individual who is carrying his vessel in a direction which is wholesome, healthy, and positively biased.

"Hence, evolution is, in this respect, a constant upgrading and qualitative improvement, incrementally, in the development of the higher sense structures. The higher consciousness is a sensing device that contains the basic instrumentation for understanding the Infinite.

"As all devices are first initiated and are qualitatively improved according to advanced knowledge attained by the individual, so is man advancing each and every time to gain a greater understanding of his own electronic instrument and his association to the cosmic force.

"This sensing instrument is recreated in an atomic manner and becomes the hardware in which man can validate and improve the operation of his higher sense mechanism!

"In such fashion did Newton and Leibnitz invent the technique of differential calculus. Topology, the higher mathematics which is evidence of a vortical shape or that space is curved, is another development. But the higher sensing understanding of those individuals who have been innovators are responsible for the present computer revolution where computer science has been able to extract large numbers and render certain equations through its high speed, and logical ability. And in other respects, optical telescopes which had their early development with Galileo, have extended into radio telescopes and radar in the present 20th century. The magnetic tape, the radio transmitter and satellite communications, etc., are evidence of man's greater awareness of electronics as a sensing device which is the hardware of atomic science to fabricate and engineer his understanding of the universe.

"These are the concepts that have been steadily advancing the knowledge of man's exterior environment, but his interior environment has *not been* advancing at the same pace. The interior environment, of course, is truly the evidence of the exterior environment and vice versa because what one can make out in his physical environment is all attendant to his ability to recognize and develop that which is in his interior mind.

"The interior mind is truly a vast computer upon which all life has been designed. Hence, it can be concluded that with all factors presently taking place with mankind in his drive to understand the nature of life, from its macrocosmic features in the nature of the formation of universes, of stars, of galaxies and in the microcosm of man himself, the least evident development in this understanding has been into the interior nucleus of man's mind.

"This is, no doubt, one of the great divisive factors in third dimensional physics, until physics becomes the true mapping and the means of determining the formation and activation of life, embracing the entire map of man. Then its contours will thus be visible to the scientists.

"The entire contour of the map is, of course, the continuity of energy from the third to the fourth dimension. Hence the proper point in the mapping of the third dimension would be from the perspective of the fourth dimension.

"In this respect, it is the limited information collated by scientists that gives them a partial view of the universe! But since man already contains the overview and knowledge to some extent, unrealizable to him as yet in his present consciousness, he can develop larger parameters to validate the movement of energy, not point to point, but rather, in a cycular manner.

"In this cycular form, the picture of life as a vortical design will be advanced, whereby science will finally come to the realization, as did Copernicus, that the earth is not the center of the solar system or of the universe, as had been conjectured and decided upon by those religiously biased peoples of the Holy Roman Church! There are other earths and solar systems that contain the balance necessary for Homo sapien life.

"Neutrinos, which are presently the object of a great deal of interest and research, have finally been discovered to have a small mass. The *cosmological constant* of the universe is not stable! The improvement, therefore, in any understanding of present ongoing research in any field of knowledge will be based upon the improvement of the individual's own sense structures. This means he will have to cross the boundary line of the very field in which he is attempting to extract information, to realize that any answer to his question can be answered in a valid and rational manner only on the basis of a logic in which energy functions interdimensionally.

"The third dimensional features of life which are the complex factors of the universe, down to the minutest feature of man on his planet and in all of the refractions of his life that he considers to be of importance, are incapable of being understood, except by understanding that they are lateral results of an initial force. It is this initial force that lives within each and every individual, that can be examined through the processes of self understanding that will alternately cross the barriers of present evaluations and provide information that will expand the individual's perspective, as he will see the evidence of his local environment from the mountain top. In this broader perspective will he gain the answer to the enigma of creation!

"What is life and how did it begin? What is my position in this process called life? Anyone who counts themselves as being intelligent would ask these questions and would realize that in the asking of these questions, he or she is the enigma!

"The enigma must therefore be pointedly resolved by entering into the resolution of the problem. The problem is therefore man himself! If any research must be done, it must be done not only from the macrocosmic or physical features of the environment that lies without man, but must be done within man himself, his inner environment. Together, there can be the recognition of the inviolate principle of life that the macrocosm and the microcosm are both functioning in an interrelated fashion and providing a matrix through which each individual, and life as a whole, takes on its orderly process.

"And yes, there is an orderly process in the ongoing evolution of the species man! The evolutionary process is the proof of an Infinite source This source is the basic principle which natural philosophers have entertained to be the causal factors of life, the Prime Mover upon which all other movements are reflected down to the smallest, minutest movement of all - the atom.

"Religion is, in all respects, a frozen system of energy constructed through a 'dam' of rhetoric and other forms of misinformation. Such static systems have reduced an individual's ability to detect the incoming signals from his own higher sensing device. Man is thus limited in the recognition of his infinite self. The political exercise of power by certain individuals who are representatives of their nation have in every respect, also been a blockade to certain degrees. Of course, this development of both religion and politics, or religion and government, which are two different perspectives, have been a recognition and a proof of the continuity of life from man's own previous life experiences. This continuity of life is constantly expressing itself.

"But the continuity of life, which is a mandate of the construct of energy, does not necessarily mean that this continuity is appearing in an upward curve in its sinusoidal movement in the interdimensional vortex. What would have been a progressive development in life, to attain a higher perspective, has not always resulted in a positive development, evolutionary-wise.

"The descendant spiral of life, is witness to man's evident technological development. This is visible evidence of the ability to construct systems which can be used to control the masses of people through fear or through satiation of consumer supplies, but has not followed with the technology of man's inner development of his higher sense structures; the visible awareness of his own mental nature with respect to its continuity into a higher spiral of the abstract Infinite.

"It is in this higher spiral where there exist *Voyagers* who have crossed the no-man's land between the third and fourth dimension and successfully reached the other side, where they need no alliance with the political expenditure of words for expedience, as a means of controlling their objective in advancing the power of self.

"Fourth dimensional physics is, to all intents and purposes, a calculus of the mind! Although it is evident in all atomic constituents of third dimensional life, it cannot be seen. It cannot be realized or abstracted by anyone unless, through his own ability to erect a picture on the screen of his mind, he can validate its existence.

"The screen of life is the entire movement of all objects that are witnessed through the physical five senses. That screen of life contains within it, only those objects which are capable of being abstracted from it, even though upon this screen is seen the evidence of all factors of the flowing river of energy.

"Because man is a developing system of signals, he is only capable of sensing audibly and visually those signals which he has already abstracted and has qualitatively improved upon within his own mentality. Therefore, an individual's consciousness is capable of determining that, what he sees in the exterior, lies within the interior of his own mentality.

"Hence, the physics of the mind is truly fourth dimensional in its nature. The study of life in the latter part of the 20th century is the recognition that what had been considered to be non-atomic, that could not be validated by electronic instrumentation, did not exist, because of the very fact that the individual researcher was not sensitive to its existence. Thousands of individuals who have advanced the existence of another dimension of the mind, evidence the existence of the continuity of life. The movement of energy through each individual's mind as thought provides him with the ability to view the nature of mind on the screen of his dimensional plasma! This is, of course, the most important development within those systems which have been given professional titles based on their systematic accumulation of information.

"These various scientific categories that have been so systematized in astronomy, particle physics, cosmology and biophysics, etc., which are features of the cell structure of the microcosm of the atomic structures, and of the macrocosm of galaxies and universes, are screens upon which the evidence of the Infinite is constantly being mirrored. So there is a fourth dimensional science whose existence is recognized by individuals who have carried its knowledge, far and wide throughout the land of earth.

"The expertise of these scientists has been gaining steady acceptance and is presently expanding the perspective of scientific research. However, since none of these individual scientists have the ability as yet to project from the mountain top to outline the features of the valley, but had only been able to determine certain features of a larger portion of the landscape, there has not yet been a complete and total description and explanation of the mechanics which make possible the mapping of life, which is presently the known boundaries of the third dimension.

"Such mapping, completed by map makers who have crossed the evolutionary boundaries of the third and even eighth dimensions, are capable of projecting into the lower theater, a picture of life which contains a higher perspective, presently the consciousness of such individuals who have graduated from the valley and have advanced to a higher perspective on a world that is far beyond the low cloud cover of an earth planet!

"Such individuals are not only cosmologists possessing an integration of all factors of life, but are fourth dimensional physicists! They have fabricated a design which is evidence of the interdimensional nature of their minds indicating the vortical nature of its design. Such an individual was Ernest L. Norman, who mapped the entire interdimensional cosmos, indicating in bas-relief, the answers to the dichotomies and enigmas of third dimensional science.

"Third dimensional physics cannot properly become fourth dimensional physics until the features of its own design are highlighted from a higher perspective. This is the unified field from which individual particles are receiving their intelligence; an intelligence that is interdimensional in nature.

"In 1956 a detailed map of the nature and function of energy and mass was published, indicating the manner and way in which Homo sapiens is a functional part of this map and an integral factor in the great electromagnetic force field movement of life. As man moves from life to life, from life to death, and from death to life in the so-called nomenclature of the physical anatomy, he is moving on a well-ordered pathway through the electromagnetic lines of force of the third dimension, and the vortical movement of energy in the fourth and higher dimensions. Mass is no longer the evidence of life and the constituent particles of the atom are no longer the viable means of holding together the instrument of an atomic body.

"Ernest L. Norman, a fourth dimensional physicist of the

highest order, assembled the evidence of the interrelationship of man and his atomic environment, indicating clearly the meaning of life. Life is a description of the interrelationship of a world of countless individuals that make up the species man, and all aligned factors which make life possible on a terrestrial planet.

"The underlying geological substrata of a planet, mineral, plant, animal vegetable, and finally man himself is an irresolute content of all mathematical equivalents which attempt to explain the forces which supposedly surround the physical anatomy. However, these forces which surround and interpenetrate the physical anatomy, such as mass, momentum, inertia, and gravity are only one factor of the forces that make the reality of life a greater understanding. Those forces are the protonic forces that function from the mind, the electronic energy field that surrounds all organic and inorganic matter. It is this field that Ernest L. Norman has mapped.

"The mind, as an interdimensional energy field force, is the basis of the *new physics*. This advance into the physics of consciousness will resolve the mystery of life, through an understanding of the cohesive nature of the electromagnetic force within man.

"Through fourth dimensional physics, man will become the captain of his destiny, because it is man himself who is the vehicle containing the navigational devices which he must learn how to operate. These navigational devices are his higher sense structures that are built into his electronic vehicle. Of course, we are not referring to the physical anatomy as the vehicle to be navigated across the ocean of the third dimension and into the unified field of the fourth dimension, another sea of energy that occupies a higher relationship with the Infinite Creative Intelligence. It is a higher sea of energy that is oscillating above the lower energy spectrum of the third dimension.

"It is evolving also and receiving the impetus of its intelligence from another sea or ocean which lies adjacent to it on a higher frequency level. However, the navigational techniques for living in the Infinite as an interdimensional being, must be developed first in the primary school; and the primary school is the third dimensional nature of life lived in physical anatomies on terrestrial planets.

"And so the rudimentary knowledge which will be the navigational devices to be applied by those individuals who are capable of conceiving the information, is the base plane upon which life will advance for these individuals. No one individual can function solely by himself, as no one particle of the electromagnetic field is capable of containing its life without its relationship with other particles in the electromagnetic field.

"Again let us restate that all particles hold their allegiance to the source of their life and that source is the whole number, which is constantly expressing its integer numbers in higher and higher relationships.

"Therefore, the advance of one individual is the advance of two or more others in two and a half times in an advancing algorithmic mode so that the future for the Earth world is a future that can be tangibly noted, because of the fact that a basic fourth dimensional physics has already been described that explains the development of navigational devices which will provide mankind with an interdimensional compass to point him to a new and greater, and truer association to the reality of his spiritual nature.

"This is the movement of life. This is the motivation of

all individuals in their movement through life, to acknowledge those factors that he knows he has within him and which will provide him with that feeling of wholeness!

"Particle physicists and cosmologists have abstracted this integrated field as the grand unified theory. Other people have called it peace of mind, stability, integration, and so on. This is the desire of all individuals to become a part of that great encircling whole which is the unification of all particles of life.

"For once that individual has attained the awareness that he is a part of the whole, he will advance his purpose in life to take on other associated factors, which will give him the recognition that he is in tune with the causal force of his life. In this manner, he is in tune with advanced intelligent peoples who are directing their Consciousness through the lens of the fourth dimension into the third dimension. It is in this theater of life in which individuals who are yet in the primary grade, are learning of the component electronic factors of their energy system.

"An infinite concept of creation must be understood, realized, and recognized as an *Infinite Creative Concept of life.* This is the whole number and is the integrated index upon which any individual is a particle of this whole number. An index can thus attain to life itself; which means the greater intelligence that is attendant upon him or her, when they have the recognition of themselves as energy fields which are transmitting and receiving devices functioning through tunable receivers and senders.

"When man realizes that he is an electronic device and that he can be 'tuned out' of a particular frequency that is broadcasting certain information, or 'tuned in' to such a station, he will then realize the dichotomy of his life is the result of being in tune with the transmitter that is the occasion for the health of himself, because he is receiving the logic and reason of information that is necessary for his present stability in his present environment; or he is out of tune and therefore his batteries have run down. He is no longer capable of finding direction in life. His directional signals have been incapacitated. This is the intent and purpose of the curriculum of fourth dimensional physics designed by Ernest L. Norman as a basic text to give researchers of life the opportunity to build their broadcasting stations with the proper frequencies.

"In other words, it is necessary to have the circuitry outlined so the individual can trace the defect that is presently causing his misalignment to the higher broadcast station from which all anterior and interior factors of life are responsible.

"In all respects then, fourth dimensional physics is a very simple proposition, a simple recognition and acceptance that all factors of life - man, plant, animal, vegetable, and mineral - are products of the higher broadcast channels. These higher broadcast channels are constantly emitting intelligent information on a higher carrier wave, on frequencies beyond the five senses of Homo sapiens. Not only must this carrier wave be recognized but the electronic knowledge of demodulating the carrier wave must be an evolutionary development from life to life.

"In the present time this knowledge of a higher carrier wave of the fourth and higher dimensions, and all evidence of the electronic means of abstracting this knowledge into the lower frequencies of the third dimensional man, is available through the teachings of Unarius, a Universal Articulate Interdimensional Understanding of Science, a definition of the Infinite!

"Since man is both the macrocosm and the microcosm, he must study and map the extensive movement of the energies which are both his macrocosmic features and his microcosmic feature. On the one hand, man has the developing ability to map the microcosmic features of his life which, up to this point, has indicated the presence of the elementary particle, the quark. He has mapped his universe and has evidenced the appearance of quasars. So therefore from quarks to quasars we have a span of great proportion which is basically the frequency emitted from the smallest index of the Infinite to the highest, up to the present time, index of the universe - the evident nature of the macrocosm of reality.

"But those two factors reside within the electronic field of each individual and qualitatively are related to him and rectified in his ability to apply this knowledge. This knowledge, up to the present time, has been applied with respect to building greater particle accelerators with different atomic resolutions, moving through the atmosphere using propulsion systems derived from petroleum products, and the ability to communicate through the means of the telephone and radio signals to communicate from outer space.

"However, these technologies are evidence only of the application of certain understandings within the realm of third dimensional physics. The technology, however, is extending the third dimensional boundary lines which is indicating that there is not only the possibility of life as man knows it on his planet but that it is probable that there are hundreds of millions of planets similar to the Earth, orbiting about their suns. Yet man has not come to the awareness of his own extension as an interdimensional factor, capable of moving through space and time, into another dimension!

"But as man is functioning to a greater degree as a developing being, to become a spiritual being, he must first complete the primary grades, the mastery of the atomic elements which are the reflection of third dimensional planetary life.

"Unarius is the resolution and proof of the continuity of life. It is the proof of the life of human beings who have crossed the great energy ocean and are now occupying life on a higher civilization. They have mastered the physics of the third dimension and are beginning to master the physics of the fourth dimension. They are now so unified in their understanding of the vortical nature of the Infinite in the movement of its Intelligence, having abstracted this Intelligence to function in a unified manner with it to the extent of their development. It is this emission of a higher Intelligence which is the basis of the libido and drive evidenced within mankind. The desire to find his home or to come back to the realization of his basic creativeness is his awareness that he was formed from a fourth dimensional energy source.

"Hence it is again, that no dimension can function by itself because they are all interlaying and overlapping and intertwining factors of the whole. As the vortex from its outer edge is indicating the nature of an immense Intelligence, each and every reduced sinusoidal movement of this energy is a reduction in frequency of the sum and total of this interdimensional vortex. Therefore, each of these lines of force contain dimensions which are interrelated with each other but on different frequency and harmonic bands. This giant whirlpool is a completely integrated association of energy structures functioning separately from each other in different frequency bands and spectrums.

"The purpose of life is to reveal to Homo sapiens, through their own infinite spark, the transmitter from which they are receiving their *libido of life.* It is at this point when man in his third dimensional perspective, becomes aware of the fourth dimension, because the fourth dimension is, as we have indicated, the transmitter, which through logic and reason must exist, because the third dimension is the product of signals of a higher potential.

"When the man of science does recognize this, his scientific research in whatever avenue, will take on much higher development, because he will be learning that he is a prototype of the infinite. He cannot possibly develop any larger map of the greater infinite, no matter in what station of life he is living if he is not aware of his connection to the various tributaries of the main river, and the main river to its final resolution to the ocean!"

Chapter 11

THE SINE WAVE PRINCIPLE OF LIFE

"If science is to be validated and not repudiated, it must be a science formulated, developed and applied within an environment in which spirit is presently making its application and its intonation. Such a science of life, states and restates the warp and woof of infinity through the life experiences of mankind, and of the sine wave principle of cause and effect!

"If science is to be the basic mechanical consideration of the continuity of the infinite in its most microcosmic reflection, it must realize the alternate part of itself in its macrocosmic reflection. Cause and effect are, in other words, a restatement of the logic and reason of the complex movement of energy from positive to negative or polarity to polarity. In another sense, cause can either be a reflection from the positive polarity or the negative polarity in which the ego of man is living, held secure within some energy body, whether in a third or fourth dimensional, or even a fifth or higher instrument of the infinite. Man is reflecting from two polarities of an electromagnetic flux field.

"If the motivating force stemmed from the positive incidence of his life, then the effects will garnish him the resultant information that had been rectified into the negative polarity. However, if the negative polarity provides the impetus, the libido and drive, then its effects will again be a manifestation, illustrating the then, present nature of the individual's consciousness. The product of infinite intelligence is, in all respects, the entire panoply of nature: mineral, vegetable, plant and animal, reflecting within their own composite natures or bodies, the effects of a variety of countless acts in which these particular factors of energy are functioning.

"There are people who consider themselves to be spiritually motivated but who denounce science. On the one hand, they consider that their spiritual motivation is a basic science, which illustrates that there is greater identity and wholeness, an integration in which man is not driven into the diminutive and the refractive nature of himself, living life through a physical anatomy and experiencing the universe inside out and upside down.

"Just because the interdimensional cosmos is being viewed through the reflective eye piece of the telescope does not mean that it does not have organization, intelligence, and continuity as a part of its association as the negative polarity of its positive nature.

"Such individuals therefore, who have had certain inner experiences and who have been aware of the reality of a higher dimension, at the same time are unsure of their footing, not capable of looking deeply within the nature of their own subconscious self. To this extent they do not countenance the so-called mechanistic view of the universe!

"The universe in its intelligence is a functioning mechanism. However, the responsibility for interpretation of this mechanism lies wholly within the understanding of interdimensional scientists. It must be recognized that science is a development from hundreds of thousands of years back, shall we say in deep time, in constellations of planets in this and other galaxies. There are trillions of humanity living and experiencing life, evolving progressively or degenerating according to their ability to analyze their experiences associated with the puzzle of life and their recognition of the alternate polarities that exist within themselves.

"It can be said that any individual who considers himself to be intelligent understands that man is not only a physical flesh and bone structure. He knows that he has had many experiences in physical anatomies, for he has been involved in many experiences in countless civilizations on countless planets throughout the galaxies.

"He must be capable of making that parallel, and realize through his own experiences that one cannot view the infinite in a lopsided manner; that all progressive understanding of humanity that is presently a part of the curriculum of terrestrial planets, was a development made possible by the assistance of more advanced, intelligent people who evolved out of the physical consciousness of an earth planet. In so evolving, they brought to the people an understanding of the positive or the fourth dimensional polarity of man - his spiritual cell structure.

"Advanced intellects have left these understandings within the texts of the various disciplines in mathematics, philosophy, physics, paleoanthropology, and throughout the literature of the land. These works contain illustrative descriptions of the interdimensional nature of life. Shakespeare, Bacon, Milton, Tennyson, and many historians, and playwrights, have indicated the continuity principle of life and the existence of future worlds!

"Man lives within his own physical body which is an energy structure living within another energy structure, the psychic body. Man himself can be considered to be a function of two oscillating energy systems expressing his continuity from life to life as an alternate movement of energy.

"In the alternate movement of the energy, there is an interdimensional relationship - one polarity functioning on a

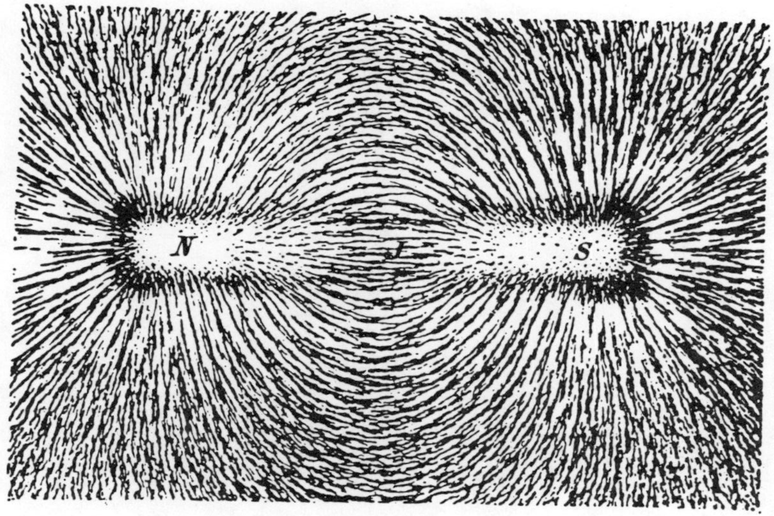

The magnetic field made visible by filings on a paper above a bar magnet.

higher frequency than the other. Benjamin Franklin coined the words positive to negative to indicate opposite fields of energy as on the bar magnet. The force of the positive pull and the negative pull on the bar magnet is equal, because the bar magnet is simply a mirror effect of the outer plasma field, the electromagnetic spectrum.

"The magnet has been so polarized that its atomic constituents have been aligned. Their north and south poles are pointing in opposite directions from each other. It is in this respect that the magnet acts as a receiver and sets up a circuit whereby the electromagnetic field refracts its oscillatory movement into the bar magnet, charging it and thereby giving it permanency. This permanency is, in fact, a closed circuit and is a *free energy generator.*

"The power, in this respect, is not necessarily in the metal itself, but in its attunability to the electromagnetic spectrum that surrounds it! In this case we see again and again that the power of all third dimensional atomic life comes from an alternate dimension, where the polarity is positive to the negative polarity of the associated atomic forms, whether it be a bar magnet or the stamen of a flower; whether it be the cell of any organism, no matter what! We see here cause and effect. This is an example of how science is lived by more positive intelligent people who have evolved to the point where they are expressing the higher intelligence of the interdimensional cosmos. In this understanding, science can now function with a greater perspective within the positive to negative oscillation of infinite intelligence.

"Man is still man when he lives without the restrictive features of a lower frequency sensing instrument (the physical anatomy). In fact, he is more a man than he was while he lived in the physical, restrictive anatomy! The reason is that he is now oscillating at a higher frequency with *infinite intelligence.* And the deployment of these frequencies is such that although they are still alternating between two fields, positive to negative, the negative field is now being recharged by a higher positive field. Previously, the oscillation was between the positive to negative frequency relationships, associated in the recreation of the physical anatomy to enable the mind to polarize the physical anatomy and charge it with the intelligence associated with the mind.

"There is stagnation when there is no movement, because without movement there is death. Movement is life. Movement is the exchange of polarities. This has been discovered by any individual of intelligence who is aware of the alternate nature of life as it deploys its intelligence into him, and is reflected in countless life experiences. In fact, it is in this manner that the causal and the effectual nature of life is finally distilled into the individual's consciousness, whereby he grows in intelligence through his own ability to masticate what are seemingly physical experiences.

"There is a mechanism involved, dear friends, a mechanism that does not take away from the *infinite intelligence.* It is truly a mechanism that is impersonal, and by which humanity learns of the greater regions of the higher Infinite. Hence, what is known as a negative experience is negative only until the individual realizes the true mechanics of life and he realizes that it is the nature of that polarity of experience, and thus changes the bias of the wave form in his recognition.

"The entire drama of life is reflected in the mechanism of the movement of energy. Therefore, there could be no life for man or for any factor of the infinite if there was no understanding of this great truth. The highest truth therefore, is that a non-progressive evolution will be experienced within each individual's spiritual cell, until he gains the important understanding that he is constantly reflecting the nature of life experiences gained from both polarities of himself - his material consciousness and his spiritual consciousness!

"In a sense, man has two homes. He spans not only continents but dimensions! He has a home on the earth, a terrestrial planet, also a home on an astral planet, and he is making journeys from life to life, visiting these homes alternately. And each time he does so, on his return visit to his astral home, he has gained some newer awareness of life and has had realizations of the nature of himself as a participant in the *cosmic movement of life.* Thus, he has raised his consciousness to so perceive and appreciate the environment in which he is living.

"When such an individual returns to his terrestrial home, he brings greater information of the spiritual nature of himself. But in this respect he improves upon his physical home by the additive knowledge that had been gained from his astral environment. It is in this alternate living and experiencing that man becomes capable of recognizing the complexity of his infinite nature through his finite and infinite life experiences on parallel universes.

"The finite is his home on a terrestrial planet. The infinite is his home on an astral planet since the real nature of evolution is the constant interrelationship between positive and negative polarities. The finite is the negative polarity and the infinite is the positive polarity, fourth dimensional to third dimensional.

"Cause and effect are factors of deep importance.

Without it there would be no life; there would be no infinite. There would be nothing to be perceived by man. Therefore, a lack of perception, a lack of consciousness, or the absence of consciousness is death - an abyss! And again, this is a tautology in that there could be no abyss, there could be no death if there was not life. We again have the validation for everything that has ever been experienced by man in his society, else his language systems would not have developed.

"Death is considered the opposite of life. Hence, we have the alternate polarity principle of positive to negative. Dark as against light. It is quite simple in some respects when one abstracts these statements into the larger reference of the reality of these word structures as they are reflecting from the true blueprint, which is the flux fields of the Infinite. We call it, the Infinite because the infinite flux field is the overall Intelligence of all flux fields, having their own individual derivative within the larger infinite field, associated in their movement frequency-wise.

"Yes, there is a great deal yet to be learned about the mind as a factor of this energy flux field, because all reflections of life are factual proofs of the existence of an infinite intelligence, functioning with purpose and direction. Each and every individual experience is an opportunity for man to recognize and prove to himself that these axioms of interdimensional science contain the principle of cause and effect.

"Science is a word describing intelligence in respect to the appearance of life in all forms interdimensionally. Science is the lens which contains the polished crystal, capable of reflecting the contour of a frequency structure of any specific electromagnetic field. In this manner, man has the reflective nature of his inner lens, a device for recognizing the higher energy structure of his mind!

"That recognition is called logic and reason. That logic and reason provides him with the understanding that nothing by itself could present an image if it did not have within it the exchange of intelligence from two opposite fields. The fact that there is not a concise, scientific understanding of the nature of the energy field in its true form, is only an indication that there have been interruptions in the development of this understanding, resulting in ignorance. These interruptions were the nature of man's past-life history.

"There are perforated throughout the evolutionary design of life and in the history of man on this Earth's plane, various and sundry reflections of the truths of life, scientifically based from a fourth dimensional physics that has been reflected in the books of the Vedas, in the teachings of Hermes, Socrates, Zoroaster, Buddha and Jesus. Many other philosophers and truth teachers have given a basic statement of the mechanics of this great energy field, the *Life Force!*

"Spinoza laid down the basic principle of the interdimensional science of life and has provided in every respect, a complete and full account of these principles, although they have not been reflected in the language of 20th century science and electronics. However, the truth of the interdimensional design of life and of the validity of man's evolutionary prerogative has been fairly well laid down by him.

"Jesus of Nazareth, himself spoke of man's spiritual nature and of his duality in the atomic body. He spoke of the energy structure of the infinite in terms of its resolution in the activities and in the deeds of man. In this respect one sees the great intelligence manifested through the consciousness of Jesus of Nazareth, indicating how man can develop his higher consciousness.

"All advances therefore, of the realization of the fourth dimensional polarity nature of life, have been refracted into the third dimension by those individuals who had already joined these two polarities within themselves, and had achieved a higher perspective of life's purpose. They had graduated from the lower schools, and were capable of functioning now in the higher energy body, the *Higher Self* or the *Superconscious.*

"Many people have called it the super ego, for it is the true spiritual cell of man that no longer is concerned with self as an individual to gain for itself its needs and to abet its insecurities. But rather, he is now concerned in a higher latitude with his relationship to the interdimensional cosmos, as a participle and a participant in the associated societies of man living in a highly polarized manner, joined as in the magnet in which the atoms are aligned and pointing towards their north and south poles.

"It is in this directional manner in which the alternate movement of energy is functioning. There is a great power when the atoms of the negative and positive poles are aligned and therefore function frequency-wise to relate to a higher energy field structure. In this respect, we give you a clue to the movement of life and how the discordant nature of third dimensional life seems so disorganized. It is only because within the experiential nature of an individual's life experiences, the north and south poles of such individuals are in disarray.

"Only some of the poles of the trillions of atoms of an individual's physical anatomy and his energy anatomy are pointed in the same direction. That is to say, the north poles of his atomic structures are pointed towards the north, and the south poles are pointed towards the south. To a greater degree, there are atoms that are not aligned towards north and south poles of his consciousness. There is no true harmony. There is a heterogeneous grouping of atoms, resulting in a lack of direction. Such an individual finds himself out of contact, out of tune with the higher frequency fields of his superconscious self, his intuitive nature, from which he is formed.

"Progressive evolution is the application of the polarity relationships of each individual's energy structure by turning the constituencies of his mind, so that they point towards the north. In this manner the individual is aligned with a higher frequency signal.

"This stronger signal carries higher intelligent information projected from frequency structures that are the sum and total of the minds of higher, advanced intelligent beings so polarized in their frequency attunements that they are receptive to even higher fields of energy from dimensions above them.

"These, of course, are dimensional quadrants of the interdimensional cosmos. In this very abstract statement, which will become more and more a part of your own understanding as you learn the principles of interdimensional physics, you will learn that you have the prerogative to rectify the nature of your own energy structure and to realize that energy is, in every respect, the atom, functioning in its alternate fashion, positive to negative.

"What is called an atom is truly a total energy structure, pure radiant energy reflected from a higher energy field. Gaining this understanding would infuse into any individual so prepared to accept these higher principles of physics, a regenerative movement of the interdimensional cosmos, to enable him to master the electronic design of his own electronic mechanism.

"To this extent, each individual has a mandate to realize that whatever manifests within his life is the product of understandings gained from his trillions of life experiences. The purpose has been to distill within him a knowledge of the energy equivalence which becomes a satisfactory balanced relationship of his life. And that means a balance of himself or herself with his relationship in interpersonal situations and the greater energy structure of life.

"Each individual's life is therefore a reflection of the ability of learning to tune his own receiver, to constantly advance the transmitter by inputting into it, new knowledge which will be a polarizing balance of life experience to be used in newer incarnations.

"This present life is therefore a reflection of the mentality of any one individual's ability to refract this scientific understanding of life. His present difficulties in life will likely be a highlight, which will give him clear insight into the interdimensional cosmos, revealing the various blocks and retrograde feelings which were manifestations of previous life experiences. With the understanding of the field equation that exists within each life experience, such an individual will become integrated and attain a clearer perception of evolution.

"These field equations of life are the loss of a loved one, losing the limb of one's physical anatomy, losing face in society. These are man's ways of learning, eventually, of the essential ingredients of his spiritual self. If these understandings are not attained and the principle behind any specific experience not resolved to give one the basis for his necessary spiritual understanding, he will regenerate the same and similar series of negative energy wave forms! These energy waves are the nature of experiential life, where one will again find, as he has returned to his brief and temporary terrestrial home, that the physical anatomy, grown through the womb, will receive similar and even regenerated factors of previous unresolved life conditions.

"Since energy, as a moving river, is never the same, it continuously adds to the nature of itself in its complimentary form, the harmonic of its associate polarity from its positive to its negative self. Each and every physical life that is lived in a new physical body finds a resurgence of unresolved or unbalanced information, information that is out-of-phase to the individual in his present environment.

"The mind, however, is not bound by a physical anatomy and therefore, can function in an alternate dimension, as it does when man is, so-called, asleep. The continuity from the past to the present, thus regenerating consciousness is the enigma and anomaly of life for man. There is only one means by which any intelligent person can find the riddle of life - through the understanding of the mechanics of his own mind.

"Hence, it stands to logic and reason that these various structures that are being examined are the reflections of a greater organism, are the reflections from the transmitter itself. Any individual furthering the quest for his spiritual identity, has no alternative but to continue this quest within the fundamental understanding of cause and effect, a basic principle of energy. How else could he perfect a mechanism and apply it for himself if he did not know of its design?

"We turn away from any allegations by any associated groups of people who would attempt to separate science from its purposeful function to determine the logic and the reason for life! To join science with spirit is the joining of the positive to the negative, within the integrated movement of the interdimensional cosmos. As we have previously indicated, understanding of the spiritual nature of life lies within the fourth dimensional equivalent of man, the superconscious self.

"It can be seen in this respect that man is constituted of atoms with a north and south pole. The north pole of man's atoms must be aligned in a direction that it is in-phase with the north pole of the electromagnetic spectrum on the fourth dimensional frequency band. The south pole is the polarity of the atom in-phase with the third dimensional frequency band. So in every respect, man is learning how to point his north pole in the direction where he will receive a higher current that will provide him with the motivating drive and libido to inseminate the spiritual equivalent that lies within this north or positive pole. To this extent, the longer he continues in the ignorance of the understanding of his fourth dimensional intelligence, the longer will he be engrossed in the separatistic nature of life and the difficulty of understanding the meaning of the evolutionary purpose of life.

"The longer he accepts the beat frequencies from his past and does not align all incidences of these frequencies which are presently resolved into a variety of 'isms' and 'osms' and systems, the longer will he cause himself even greater frustrations, because he is 'out-of-tune' with the infinite nature of himself, and as a result is receiving static.

"So in this respect, man is turning the channel of his mind in between two stations, and it is not entirely focused on either one or the other. He is pointing his channel selector on the lower channel and receiving the static of that channel from the basic emotional forces of his past, containing a medley of signals of a great staccato of negative information.

"On the other hand, he is attempting to move that channel selector to get a clearer signal, but he is incapable of doing so until he learns the mechanics of this electronic principle. Therefore, in attempting to interpret the signals, he is beset by a variety of hypotheses and theories as to the nature of the static energy third dimensionally, All that is needed, of course, is to know of the design of this electronic receiver and thereby tune into the channel from which he would receive the information for a complete understanding of himself. The higher channel, which is the station of his own superconscious self, will reveal to him the answers to all problems and enigmas, which are products of his past from his third dimensional perspective of life.

"The fourth dimension is the spiritual home where the resolution of life has been so reformed. It is the unified state of consciousness, the grand unified field equation. This is the principle that lives within each and every individual and is being attained in life's development as a student of the Infinite.

"The mind is a mechanism that operates on the principles of polarity relationship and polarization, positive or negative. These polarity relationships indicate the basic nature of life in which the factors of life experiences are becoming integrated to provide a clearer picture of his nature over many thousands of lifetimes. But if the picture is unclear and is out of focus, the energy structure of that individual has not been polarized from his previous life experiences. It is as simple as that. And as complicated as it may seem, it is realized by any dedicated student of the *Science of Life,* that he contains within him the ability to reconstruct his life to reflect the higher components of a higher frequency, which will affect the reality of himself.

"When it is quoted that such and such an individual has a clear mind, its meaning is that the individual has adjudicated the essence of the complexity of life situations, covering the specific aspect of interpersonal relationships. Such a mind has attained certain realizations which have moved the understanding of that individual to a higher statement of progressive evolution!

"To possess a 'clear mind' means that the individual has gained access to his computer, and as a result has gained greater information of himself that can be applied towards a certain set of positive constructs. These constructs provide the means for polarizing the individual relationship with other persons thus strengthening the consciousness of each individual through the association of each to the other.

"Each individual man, is therefore a component part of a mechanism. As man is an atom of the Infinite, he is a component part of the infinite. That individual component part in whatever position it takes as a transistor, rectifier, resistor or capacitor gains its intelligence only in terms of its association to other component parts. Therefore, the integrated nature of man as a factor of the infinite, is one that is attained by his association to the society of man through his realization of the principle of attunement.

"If one individual component of an instrument is attuned to the higher polarity field, that power is instantly projected to his alternate polarity, to the negative side of itself or the south pole, therefore, regenerating it to all other associated atoms and expanding its force for positive work. However, if the other component parts of the instrument do not attune themselves, do not point their north poles to the source of the power on a higher energy field, there will be a reduction in power. Eventually this higher power will be expended, because there is no exchange from the other polarity of this mechanism, and as a result, the depletion of power is inevitable. This is a fact of great historical insight, in that there have been many positively powered individuals whose physical anatomies were detuned because of the lack of a higher polarity relationship with another individual or individuals in the society!

"In the late 1900's the concept of alternate current, which is now the basic principle behind the function of all motors was a result of Nikola Tesla's infinite understanding. His mind was in tune with the fourth and higher dimensions because he is a product of the higher field equation of the infinite, and was therefore capable of carrying this power. The same mechanic was used in Niagara Falls, in great transformers, to implement and to add to the electronic revolutionary development of the 20th century.

"Nikola Tesla was in tune with his superconscious self while so associated in the negative polarity of a physical anatomy. And it was in this attunement that the entire mechanics of the movement of energy was so realized by him in his conscious mind, which was held within the third dimensional negative field of a terrestrial planet.

"This is one of many examples of the continuity principle of life, the sine wave principle which expresses this continuity. The cause and effect nature of this sine wave principle seen in the acts of man as he lives in two adjacent dimensions is reflecting the nature of his understanding from one to the other, thus adding or subtracting according to his ability to recognize the principle so associated in this, the intelligent equation of infinite consciousness!

"Yes, Infinite Consciousness is the objective of anyone's

progression! It is the direction towards which man must be pointed to gain an integrated consciousness. And to gain infinite consciousness is to turn the north pole of the electromagnetic field towards the incidences which reflect this infinite consciousness, a superconscious circuit which is the higher self of man, which does direct the lower self the lower sense body, which is recreated through the subconscious portion of the electromagnetic body, the grand computer!

"The equation therefore, for an intelligent life of balanced integration, wherein the picture of life is untarnished, will be gained, not by fleeing from the past and shutting down the picture that flashes into one's consciousness. But instead, to realize that this is the lower field, the negative polarity that has not been polarized and is therefore the product of this equation of the movement of energy from its higher to its lower field. Man's purpose is to gain insight into the nature of the breakdown that the present effects of life are indicating as an unpolarized factor that has not been rectified.

"Study the energy principles of this Unarius Science and your life shall gain a great clearness, and the tarnish that you have allowed to accumulate in the dusty attics of your lower mind will, as a result, be cleared away. The psychological factors therefore, which are products of this cleaning, offer other, deeper realizations. Although this has been given many names in the vocabulary of religion, metaphysics and mysticism, it is a purely scientific derivative of the cause of any situation presently out-of-phase and considered to be a malfunction in the individual's life.

"Yes, there will be remorse, but it is not occasioned by the devil. It is not occasioned by any fabrication or composed picture which would provide the alibi for such remorse. It will be indicative of a recognition of the exact nature of the misuse of energy. It is such ignorance of the dynamics of energy that has resulted in the malformation of the energy structures of both the aggressor and his victim.

"The universal principles of this interdimensional science are the medium through which all of the dust and corrosion, occasioned by the rust of an unused mind, can be rectified. To be a spiritual being, in short, is to have an active mind that is functioning at all times interdimensionally, that has not been shut down through the inadequate knowledge of its mechanism."

Chapter 12

THE FIELD EQUATION OF INNER SPACE

"The controversy continues. The battle rages; East is touching West! The mysticism of the East is involving itself in the logical positivism of the West in the denotation of a fundamental physics that had been a development of Western empirical scientists. The West is therefore the sum and total of an advancing technology, which as a whole, has taken into consideration the formulated hypotheses and their application as theories, into the practice of a high technology in the industrial societies of the west. The West has crossed the barriers of Europe into the Soviet Union, into Asia and China, and has gained entry to India where mysticism has had its greatest emphasis.

"Even India is not untouched by western empiricism. The experience of trial and error in research provides scientists with the means of moving man in a direction so that he can begin to master the elements and develop instrumentation that will apply the electromagnetic principles of energy which he has been formulating since the time of Kepler.

"The controversy is therefore a controversy not necessarily between East and West, as a physical statement related to the map of the earth, of its hemispheres of west and east. Truly, the controversy has to do with the prescribed nature of religions and philosophical fields of inquiry into the nature of life. These religious expressions have been given different labels, as Mohammedanism, Buddhism, Christianity, and many others which are products of the combination of Zen, Hindu anthropology, Buddhistic, Taoistic, and other relationships, factors of the basic, fundamental statements of life as an inner and an outer reality.

"The controversy in present-day science is in the attempt to understand the basic quantum of the atom. Many scientists have been mentally fortified with certain spiritual factors which have been brought from the east. They have been students of the teachings of Buddhism, the Tao, Hindu, Christianity, Zoroastrianism and Judaism, etc. We see at the present time, the formulated empirical constructs which have been the basis of the development of physics since the time of Aristotle. Great particle accelerators smash open the innards of the atom, in the attempt to reveal the enigma if atomic energy by examining the elementary particles: hadrons, quarks, etc.

"Some physicists have used the term, 'Looking Glass Universe'! They are closer to the truth in this synonym, because the mirror effect that is the present atomic picture of life is, in every respect, the reflection from a looking glass or mirror. Man is seeing the reflections of his fourth dimensional body while he is viewing optically his third dimensional environment. In this manner he can attain the mental orientation of his real image, because his third dimensional image is not yet complete in the coefficients, which are at the present time, giving him the constituents of himself as an energy system.

"The true picture of life will then be realizable, because as an energy configuration, he is not only the *Looking Glass,* but he is the reflection of the *Looking Glass*. He is the transmitter that is projecting the image onto the mirror, and he is the demodulated image, atomically speaking. In every respect, statements now being made by members of the scientific community: microbiologists, chemists, particle physicists, etc., probing into the constituents of stars and planets and other features of the universe are reflecting new light from some area in deep galactic space, and from the deep space of the atom.

"It is no wonder there is great consternation in these investigations into the nature of the atom and the universe, which have ended in a blind alley. The reason is because the stone wall that still remains in the investigation into the nature of the electromagnetic flux field of the atom, is man himself! This statement is close to the truth, because all investigations into the anomaly of the third dimensional light spectrum take man back to the reflected nature of third dimensional life experience. In other words, the conundrum of research within the equations of the light, from within the radiant light field of the third dimension, provides little possibility of determining anything other than that which is reflected in this atomic spectrum. One cannot get outside of the light coefficient of a dimension. The coefficients of all intelligent life are existent within the light frequency of the oscillatory movement of *radiant energy,* which in its highest formulation, is not mass. It is pure energy!

"The coefficiency then of the radiant spectrum of the fourth dimension is the basic proposition that has been set forth previously, and is the boundary line of the third dimension. All investigations into this field of higher energy spectrums can not be accomplished with present instrumentation which measures the various coefficients of momentum, inertia and gravity by using the symbolic logic of third dimensional physics.

"It will take a new physics to measure the *radiant nature of the mind*, as it does exist beyond the third dimensional light barrier. For this reason, the mind is a deep mystery to physicists, neurologists, psychologists, psychiatrists, and other medical practitioners, because it is a flux field, an energy system, which functions on a fourth dimensional light field of radiant light.

"The great anomaly then, is that the radiant light emitted from the solar disc of this and other solar systems, is a compaction from great energy force fields of the fourth and higher dimensions, which are beyond all possible conscious conception. Call it *cosmic consciousness* or the *cosmic mind.* It is a fourth dimensional energy field that is functioning within the nature of its own consciousness, from its positive to its negative polarity. Therefore, 'cosmic intelligence' is oscillating into its demodulated negative form which is the sum and total of all that It is. The atom is the smallest quotient, the lowest frequency in the centripetal creative intelligence of the infinite!

"This is the Force presently being lived and experienced by all Homo sapiens in their atomic form. This nuclear compacted energy is also projecting out into new and different relationships through its expanding negative polarity into another infinite. This is the *looking glass* reference in which all things which were, are and will ever be, as radiant energy are now regenerating themselves into a new perspective as dimensions within the great *interdimensional vortex* of infinity.

"Hence, all investigations into the *Cosmic Egg,* if we can use the term, means that the investigators must transcend the reflected nature of light, as it presently is the reformed image of all attendant features of atomic life. The only way that investigation can be accomplished is to cross a bridge, or in another sense, climb a wall and peer over it. The wall and the bridge are, in every respect, the electromagnetic flux fields which are presently the existence of the third dimensional life!

"In other words, man can obtain the answer to the perennial puzzle: 'Which came first, the chicken or the egg' when he begins to realize that he is both the chicken and the egg, the cause and the effect. This then contains the elements for the unified field equation!

"Fermat's last theorem presents an anomaly that has not been answered, because of the nature of the question. Man is the egg and he is also the chicken. Until he does develop his higher intelligence he will peck away, satisfying himself only with small grains of information, but at no time will he ever be capable of satiating himself sufficiently in his need to cast his eyes over the wall of the atomic reflection of life to learn of the whole number!

"Interdimensional physics is the continuity of all research, which has as its primary purpose, the understanding of the *whole self*. Man is the real subject of research, and is the product of research! Whatever reflections of atomic life seen as tornados, hurricanes, earthquakes, and volcanos, are the nature of man, himself, as he is experiencing life on the terrestrial planets, which are in every and all respects, the learning of the anomalies that exist within the energy field of mass.

"The psychic anatomy then, is the true research project in all factors that have to do with the understandings of the stresses and strains within the substratum of a planet and also its climatic factors. Understanding the stress factors of the planet will explain the ratio and relationship of the energy fields of the electromagnetic body of man.

"We had indicated previously that the electromagnetic body, the electronic man, exists in three quadrants, and that these three quadrants or circuits are the basis from which all electronic instruments have been created or manufactured. The circuit fields of the electronic body contain information that validates the existence of the energy body of man, and explains it's structure and function. This knowledge will also provide the information to rectify all societal ills and other problems associated with the integration of the physical anatomy and the mind of man!

"The associated functioning of a healthy body and a healthy mind will result in a healthy and vigorous society, a healthy and vigorous civilization, and a healthy and vigorous planet. And as the cells of each of these areas become healthy, vibrant and whole, so will planets take on their basic position in the cellular structures of the galaxy.

"As a result, the galaxies become a greater cell and a vigorous part of the greater anatomy of the universe. Then it can be seen that the *design of life*, which was interpolated by and investigated by certain previous philosophers, is a great organism in which all of its parts are cells; each cell receiving the substance of its growth and of its positive regeneration from the *interior* of its own energy system!

"The growth of an individual cell demonstrates the very complex nature of energy, and can provide a great insight into the internal mechanisms of infinite intelligence. Yes, one tiny cell, of the hundreds of billions of cells in the physical anatomy, is proof of the very complex mechanics which are the functioning parts of the *interdimensional vortex*, a Venturi Tube, diagraming the field forces of life.

"This spiritual anatomy is a viable mechanism, a reflection of the true protonic cell, which is the *cosmic mind* or the cosmic egg. Cosmic intelligence is complete, abstract, and unknowable by man living in the *looking glass universe*, but becomes a developmental knowing, through each of the cellular structures of man! The greater the knowledge of self; the greater the knowledge of the origin of self. The *origin of self* is within the nature of the cell, itself. And that cell we speak of and refer to, is the *electronic body*. the spiritual anatomy. This is the true cell!

"Since this cell is the protonic egg or the cosmic egg, in every respect, it contains the answers which can be resolved through life experience, where the equations of its movement become a factor of the individual man. That factor becomes his through knowledge and wisdom. Wisdom becomes his basic lens providing him with the ability to see. That seeing is his consciousness. It is a consciousness, fabricated not on the lower spectral emission of the light of a third dimension, but a consciousness reflected from the Light of the Interdimensional Cosmos.

"The entire interdimensional cosmos, in which man lives and receives his experience from life to life, dimensionally speaking, is the structure of the interdimensional Cosmic Mind. Knowing about these interdimensional principles that have been the Philosopher's Stone from the beginnings, of so-called time, is being a part of the wonderment of what is the true and the beautiful, that never changes but always progresses.

"That beauty, and that abstract truth is attained in successive lifetimes, and is, at the present time, being formulated by many individuals involved in determining the scientific equations of life. But many are still carrying 'coals from Newcastle', because they are reliving their past lifetimes when they were themselves students or disciples of either Pythagoras, Plato, Aristotle, Anaximander, Hermes, and Anaxagoras, and a host of other Grecian and Egyptian philosophers.

"Many such individuals are mystics. They believe there is a life out there somewhere, and that their energy does not die after the physical body stops in its mechanical movement. Of course, that is the intelligent approach. Energy is the nature of, and is the reflection of Infinite Intelligence. However, it is one thing to conjecture and another thing to develop a concise principle that will diagram this energy as it is the moving force of each individual's mind, and a resolution of his electronic body.

"That diagram has not yet been developed by engineers in electronics, microbiology, biochemistry, nuclear physics, or other physicists and philosophers functioning within their respective disciplines. It cannot be accomplished on the basis of a physical science, because of the great quandary that has been discussed in the previous paragraphs.

"Only a scientist who functions from the fourth and higher dimensional quadrants of this great interdimensional movement of energy, is capable of resolving the enigma of what exists and how it exists beyond the speed of light, and to properly resolve the motion. of energy into a universal statement, a universal physics.

"Interdimensional physics is a basic description of the manner and way in which energy functions in the fourth dimension. This is the attendant movement of energy, which is the nature of the intelligence of man as he exists within the nature of an intelligence field in which he lives beyond the oscillatory movement of third dimensional light.

"This energy movement cannot be measured with third dimensional instruments as it does not contain mass. Therefore, a new mathematics is needed. This which was the opinion of several physicists, including Robert Oppenheimer, Erwin Schrodinger, and others. The proposition that a description of the physics of the fourth dimension as a realizable factor, is contained within the dialogues from such fourth dimensional physicists, who presently live in a higher quadrant of the Infinite and are invisible to the average third dimensional man. But that invisibility is only a function of the frequency in which the physicists or scientists or whoever is being referred to, lives.

"Therefore, any individual who would attempt to solve the ambiguity of the language of third dimensional physics and attempt to formulate an intelligent equation as to the continuity of life within the framework of third dimensional energy forms, would by necessity, need to have developed in his own mind, the mentality, which would refract and reflect to him the information being charged and communicated from such advanced, intelligent Beings, whether he be a political scientist or whether he be a physical scientist.

"Such advanced individuals have communicated these basic principles of the interdimensional movement of life as it exists in the higher frequency plane where they live and are much more resilient in their intelligence. They are much more alive in this intelligence than are the people who live in the lower levels or frequencies of the Infinite. The vast majority of people on earth do not have that development or capability at this time. It is not that they will not! For man's mandate is a progressive evolutionary purpose of life, to always update information and move ahead within the great nature of the Body of the Infinite to enjoy the presence of a higher knowledge and understanding of self.

"The great debacle within the scientific community will lead to the resolution that it is the development of one's mental attributes that has been blocking one from further progress in unlocking the enigma of the energy fields of life.

This will come about when the scientist recognizes that the interior of each individual's mind is a functional whole. Science, then, will be realized as the derivative of all reflections of life from the microcosms to the macrocosm.

"In 1956, the interdimensional scientist, Ernest L. Norman, using the advanced psychodynamics of which we are relating, communicated these fourth dimensional principles of science diagrammatically, in a series of 20 dissertations.

"The basic proposition of these dissertations provide the answers to the theorems of all science: in biology, anthropology, sociology, physics, cosmology, psychology and philosophy. Man does now have the opportunity to begin that forward movement, which is the *continuity principle* of life, the expressive nature of energy as it functions in this very complex electromagnetic flux field, the energy within which all mankind lives.

"Ernest L. Norman, therefore, proves the existence of life as a scientific relationship, the existence of life forms as an evolutionary regeneration from the fourth and higher dimensions, where this energy has been rectified as Homo sapiens to be regenerated into new relationships for the future of all mankind now living in his third dimensional anatomy.

"The entire picture of life is not complete until it is understood as an Infinite Creative Intelligence, moving from a centrifugal to a centripetal polarity; and in this oscillation from its positive to its negative polarity, exchanging the intelligence of itself in the expansion and the contraction of itself. Man thus gains the entirety of his consciousness, always expanding into new formulations in this exchange of polarities.

"Therefore, as man himself is a cell of this greater infinite, he also expands and contracts within this great movement of energy. So man is moving between his dual polarities and learning of himself as he lives in the contracted part of his energy field, in an atomic body, which is the sum and substance of the compacted nature of infinite intelligence.

"But man also lives in the expanded state, his fourth dimensional centrifugal station of life where he views the infinite. In this expanded fashion, he no longer feels the compression of a solidified, so-called, atomic body and all the attendant factors that are a product of this compaction, which were limitations in his understanding of infinite intelligence.

"In this exchange, therefore, between his fourth dimensional and third dimensional polarity relationship, he gains a higher intelligence from life to life, providing himself with newer perspectives and recognitions of his placement within the Infinite Cosmos.

"Scientists must therefore realize, that in their pursuance of the answers to the nature of the *looking glass universe* they are seeing themselves, reflected in their physical cell bodies as well as in the nature of the perplexities of their mind. This is true whether they look into the cell, into the deep sea of the atom or into the deep sea of outer space. Hence, all mankind, scientists particularly, must take themselves in hand and realize that they are the *missing link!*

"The missing link is found in the recognition of the impact upon man of the continuity of his energy from his own previous lifetimes. This oscillating energy has set up systems of positive attitudes of mind, or negative attitudes of the mind, blocks based on various traumas and other types of ego developments, which have mired such individuals within the atomic mass from which he is attempting to escape. There is no escape, my friends, except through the recognition of the nature of the protonic force that lives beyond the eye of the needle. Subatomic research will not be resolved until the protonic forces are recognized as being the determining factor of its existence.

"So let us look into the mechanism of energy as the field equation of inner space, which provides the evidence of intelligence and an understanding of life. The understanding of the principles of energy associated from the past-life experiences of an individual will resolve his present disparity and difficulty in formulating propositions which contain the logic and reason of an interdimensional and progressive consciousness.

"The mechanics of the brain cannot be easily or fully understood until we comprehend the dynamics of the Mind. The entire and central principle underlying all life is the principle of polarity; the dynamic oscillation between two opposite force fields that are operating on essentially two different frequency bands, one higher than the other. It is in this alternate movement of the current, that is being exchanged or oscillated along this carrier wave, that all intelligence is so concluded. In fact, the entire prescription of life is a relative factor in this dynamic duality of the movement of energy.

"In this respect, there can be no resolution on any field in which man lives, whether it is in the lower oscillating field of the third dimension, or on the fourth or higher dimensions, that will ever give him a clear, concise understanding of the nature of infinite intelligence! That nature is prescribed within the dynamic energy envelope in which man functions. That dynamic energy envelope is therefore seen to be a containment within its sheath or field of two force fields, two electromagnetic fields that are moving in and out of each other, so to speak.

"In this fashion, as we have constantly reiterated, as a means of enlarging your perspective, that the interdimensional cosmos is a breathing in and breathing out of two adjoining fields of energy, the centripetal and the centrifugal force fields. This breathing in and breathing out is similar in some respects (in a crude fashion) to an accordion or a wire that has been twisted in a circular configuration and which can be pulled out to its maximum and then allowed to contract or spring back.

"Man, you see, functions in this fashion, and therefore, the mechanics of this movement must be determined by the very quotients of man's makeup, which are vortices of infinite energy; vortices that contain the entire history of any individual. The mind is the fingerprint of man's consciousness regenerated from the conception of his evolutionary beginning!

"Within the infinite nature of Infinity there is no beginning, nor is there an end. It is only a philosophical statement that such a beginning would be the recognition of that spiritual cell, which is man's origins. Therefore, that beginning is the awakening of consciousness for mankind in his present life.

"Now we are back, to the basic fundamentals from which all rational concepts of life can be understood. The logic and reason for the dichotomy that is seemingly a chaotic association of life experiences, then becomes quite recognizable as a universe of great order!

"This order and plan is the entire nature of experiential life for the lower development of man during which man is learning of the underlying and overlying nature of his own energy field. So we take you to the diagram on the computer or on your blackboard, and indicate primarily that the polarity nature of mankind will reveal the essence of his intelligence, both of his lower and of his higher nature. The physical anatomy is the basis for all present-day research in microbiology, biology, and all of the allied fields involved, such as chemistry, and in the fields of health, where medical knowledge is being applied to the nature and the function of this anatomy.

"The physical anatomy is the visible appearance of man, as a viable organism that contains a life force. It is the negative polarity of this field force and is revealed through the sense organs and through other types of instrumentation that amplify the five sense organs. The entire gamut of research into the physical anatomy has indicated a variety of interesting facts that have been garnered over the several thousands of years of history known to man, but more specifically, since the time of Harvey.

"The nature of the chromosomes and genes reveal that they are functioning within the dimensions of the large proteid molecule, where the inner structure of this molecule indicates the great intelligence which directs the function of the cellular structures of the anatomy. The DNA code, has been, so-called, cracked, and the secrets thereby, of the interior of the molecule, have been revealed.

"The brain, in many cases, still accepted to be the seat of the mind, has been investigated, and to some degree, has been mapped. Within the organ of the brain certain understandings have been revealed of the left and right hemispheres, how they function to activate the physical anatomy itself as well as the movement of the cardiovascular system and the pulmonary system, etc.

"Certain information has been obtained of the nature of the cortex, the frontal lobes of the brain, which have to do with the gestation of reasoning, the ability to deal with abstract numbers or abstractions that can be formulated, etc. At this point we reach the present edge of the investigation. It has been quite definitively indicated that the motor functions of the physical anatomy are activated by the rear lobes of the brain and in different portions of these lobes. In the functioning of the mechanical factors of the body, there is sufficient understanding of the electrical charge and discharge circuitry that is responsible for the movement of the physical limbs, etc., however little is known of the psychodynamics of the brain.

"In fact, nothing truly is truly understood of the ability of the individual to carry within himself certain abstract understandings that are invisible and are not indicated on instrumentation. Man is constantly deriving certain quotients from abstractions of his mind, through a complex equation that is carried within himself from his life-to-life experiences. These are rectified in the movement of the entire organism of his physical anatomy.

"The brain has been investigated to the extent of its application to the physical anatomy as a motivating force. Yet there is little knowledge as to the gamut of complex relationships that indicate the motivating force that provides for all associated factors that give this physical facility its basis and its foundation.

"The young science of parapsychology, has made some steps forward by indicating the invisible field that provides the motivating force for such factors of man's life, which are not capable of being rectified through the behavior of his physical anatomy. Parapsychology deals with the paranormal factors in the physiology of the brain and the physical anatomy.

"However, parapsychology is itself, bound within the nature of its own investigation, because it has not taken into consideration, or has not yet seen the full nature of the map of the energy structure, which is responsible for those paranormal factors, such as psychokinesis, telepathy, out of the body experiences, existing on a different level of understanding.

"There are, of course, thousands and hundreds of thousands of examples that have been accumulated from the life experiences of many, many individuals whose experiences have been collated providing evidence that there is *something out there* beyond the frame of reference of time and space and of the concrete reality of the atomic world!

"There are in the literature of parapsychology and in literature of the paranormal and metaphysics, experiences of many people that have led to a certain development and systemization of knowledge, which would provide some answers and clues to the probability that Homo sapiens lives continuously after the completion of his cycle in the third dimension. That is to say, he lives after the physical body has lost its power, and the atomic constituents of its cellular structure are no longer operative. So there is witness in the long history of man that there is another world or worlds on which man lives as Spirit.

"However, within the written annals of man, during the past 10,000 years, there has been no physics. There has been no written evidence; not in the literature of science or philosophy nor in any other writings with the exception of the Unarius Science of Life, of any proof of the higher mathematics of man's mind as it functions in a plasma that provides intelligent interaction with the Cosmos.

"This is the purpose of this book, to put forth the full nature of the *fundamental dynamics of energy*. For life is not a mystical representation. It is not supernatural or supernormal. These expressions are only evidence of the limitations of the investigator and of those individuals who have accepted the primitive signature of such investigators, who have stated, 'This is the way it is and anything else is beyond any possible understanding; it must be supernatural because it is not natural!' The vocabulary describing the unknown has been developed and embedded with the ignorance of people and forces who would wish to keep man in ignorance so that they could take advantage of this ignorance. The factors involved in such ignorance are lack of knowledge of the *real* nature of the man himself as a participle of the infinite creative intelligence.

"Now is the appropriate time and cycle of evolutionary life for earth man to have the basic fourth dimensional physics, diagrammed to describe factually what goes on in the invisible world - the world, invisible only to the five senses, of course.

"Because there has been a development evolutionary-wise, there are more and more individuals who are able, through their own logical and reasonable attitudes, to become aware that there must be something out there. We are referring to the macrocosm and the microcosm, to man's own nature and to the outer appearance of this nature, in nature itself, to the Cosmos; that he is now more capable of accepting and being able to deal with the information that is now presently being demonstrated to him in the library of an Interdimensional Science.

"The Unarius Science is such a library that has been communicated in an interdimensional manner, from the libraries of the universities in these invisible worlds, where they have been created as electronic writing, so to speak, These libraries are contained not in one building, not on one planet, but are maintained on an array of planets that function on a higher frequency of Infinite Intelligence!

"There are libraries of information that have been accumulated from the life experiences of advanced, intelligent people who now live in non-atomic bodies in an electronic energy configuration. It is from this library on Eros, one of the higher dimensional planets, that this science library has been communicated electronically. along the carrier wave of the mind that is the connecting link between each dimension.

"The Unarius Academy of Science is the product of the endeavors of men, who as Brothers, live both on the higher frequency planes and on the lower frequency planes. They have made possible the third dimensional expression of this interdimensional physics by demodulating such information from the higher carrier wave into the lower frequencies of sound and light waves that are the matrix of the third dimension.

"This library of the physics of life was initiated 35 years ago through the auspices of Dr. Ernest Norman and Ruth Norman. Because of their interdimensional origins and ability to tune in, frequency-wise, to the fourth dimensional and higher libraries of the advanced civilizations that exist in the higher frequency worlds, they have rectified this information to bring it into the perspective where all man can learn of this interdimensional Science.

"The present Unarius Library is a developmental and an

extensive history of the past of man. It details an understanding of the origins of man himself as a spiritual cell, describes the existence of these spiritual cells as they have evolved to higher recognizable intelligent receptacles of the Infinite, as well as it indicates the evolutionary nature of this development. This evolutionary nature is the alternate movement of the energy of the infinite as it is abstracted and lives in the mineral, vegetable, animal, plant, and of course, in man as Homo sapiens.

"It is important that this understanding be related at this time so the evidence of its nature will be recognized. The question is always asked of anyone who has familiarity with these interdimensional principles: 'Where on earth did you learn of this information? And how is it that you know of this and I don't?' - evidence again of the diagrammatical factors of certain ego deflations, of discovering answers to the innumerable anomalies and enigmas that have been factors of third dimensional life.

"Such an individual is not, in himself, the only person who is able to speak coherently about the wholeness of life. He has tapped into a Universal Library, and that library is completely impersonal and knows no ego. That is to say, it has no relativity to any one individual. It is a property of the Infinite. It is the accumulated nature of life experiences of hundreds of billions of trillions of peoples, who as a whole, validate the entire movement of the infinite, that is presently the acceptable feature in only one of its great sine wave movements in its energy makeup!

"Anyone who has the understanding that 'there is something out there', and is beginning to properly rectify this understanding within the nature of a viable or practical application, then has developed certain abilities, electronically-wise, to be able to rectify the higher signal that exists in the higher circuits of himself. Therefore, he has ded-veloped to a certain degree, a greater number of circuits which are imprinted within this higher superconscious self or circuit, and is capable of access to it. This provides him with a higher degree of awareness.

"Such an individual who has developed Cosmic Consciousness is sensitive to a higher frequency band and is a cosmic visionary. Such individuals make up the receivers on third dimensional worlds. for they are the ones helping to move man forward by making available knowledge of the blueprint of man's energy system.

"However, we have digressed to a certain extent and will continue and take up the relative relationship between the physical anatomy and its motivating force, the transformer mechanism which is the brain. The brain can therefore seem to be an organism that is functioning electronically as a switching device. A complex series of signals are relayed through the various cell structures of its organisms to a specific organ of the physical anatomy, making possible for the anatomy to transmit signals that are sent to it.

"The complexity of this mechanism is known to some extent, but the lacking knowledge of it, is its relationship to the master switcher, the master brain, the master computer; because the brain is only a terminal of this *Master Computer.* So let us readily see that the physical brain is a terminal that is so connected to the master computer that it will carry the necessary information to the work at hand. The terminal provides access, but the true nature of the terminal is still not complete until the realization from whence it receives its signal is attained.

"This is the Master Computer, designed by Infinite Intelligence, existing in an adjoining dimension beyond the physical sense receptors of the physical anatomy. It is unknown, except from the evidence that is gained through the terminal as to its existence and the anomalies and enigmas in which present life experiences express to each and every individual, as a continuity of previous life experiences!

"The previous life experiences presently being received into the terminal of the brain, or the switching organism, are being carried as electronic signals to the physical anatomy. Many of these signals from previous lifetimes do not carry an intelligent continuity in terms of an integrated relationship to present life situations. And it is in this disparity where it is an out-of-phase association with what has occurred in a previous lifetime and what is presently taking place, that lacks any coherence, logic and reason. Such may be the fact of an individual who wakes up from sleep to find himself covered with a welter of red marks over his body! As there has been no instance of a biological or any other factor previous to this time of an instance of such red welt, there is no organic basis for any investigation into the cause. Hence, this dichotomy is an anomaly and an enigma, for it is an unknown factor!

"It is slowly coming to the recognition of the investigators in the biological and the psychological sciences, that all evidence of any disease whatever, that causes a break in the integration of consciousness, is a resolution of some past association of traumatic life experiences. Therefore, the evidence of the map that provides an understanding of the physics of these mental and physical breakdowns in life: malfunctions, diseases, whatever, is the next step and the next development in the evolutionary education of mankind!

"The existence of the mind as a viable organism that is

both functioning in the physical anatomy through access of its terminal, the brain, is also functioning in an infinite manner to provide evidence of its interdimensional nature. It is just because the mind is an energy field, which is interdimensional in its construction, that it must be known so that each person can facilitate his progressive evolution.

"The curriculum of this progressive development, which is the on-going present nature of life in the latter part of the 20th century, will be the application of these interdimensional principles of physics, which principles cannot be operative if man does not recognize the specific nature of his mind. The existence of the mind is a factor in which all people agree, but as to its nature and function, there is generally disagreement.

Chapter 13

THE PHYSICS OF THE HIGHER DIMENSIONS

"Interdimensional science provides the answer to the entire fabric of not only the many fallacies that have been entertained as a hypothesis to explain the enigmas of life lived by all human beings, but it posits to a greater degree the enigma of science, itself. Metaphysics, as a higher statement of the nature of life in the invisible higher dimensions has not as yet been capable of describing the principles of interdimensional physics.

"The tool which will integrate and synthesize the many factions that are presently functioning within the various disciplines, is portrayed under the banner of science. The old hyperbole of wedding science and religion as a means of giving a complete whole understanding of life, has many problems, in that science functions with its present concern with the nature of the movement of the energy of the atom, and religion has to do with the movement of man's mind with the understanding of its continuity after the barrier of the atom body has been released.

"The nature of this 'Force' known by many names, as spirit, god, the infinite, has not been diagramed, at least to the extent that it can be accepted as a factor or as a common denominator, with which all individuals can work; a diagnostic tool, so to speak. Mathematical and theoretical physics are the diagnostic tools of science, including the experimental procedures.

"However, mathematics, as a symbolic logic, does not function in the extension of an interdimensional synthesis that becomes the joined products of third and fourth dimensional physics, as an interdimensional science of energy. Although religious institutions do contain certain rudimentary statements of the ongoing and embracing movement of intelligence, which spans the visible sphere of life, the atom, and the invisible sphere of life, spirit. Religion is not a real factor, as there have been incorporated into its belief system great misrepresentations, causing additional misunderstandings of the reality of consciousness.

"Science provides the basis of a logic and a reason and an analytical technique. It is the analytical technique that is the evolutionary developmental means for mankind to embrace principles of logic and reason. Nowhere in religion, mysticism, metaphysics, or spiritualism is there taught a physics of the inner life, a physics that spans two dimensions. For physics is an evolutionary development of individuals and is the basis for knowledge of the world of the nonatomic sense structures. Such individuals have been considered mystics. A mystic is one who has communicated with the unknowable world, at least in the terms of present third dimensional physics.

"If the mystics and the metaphysicians speak of the nature of the abstract Infinite in terms of the continuity of consciousness and that man resides on a higher plateau of this Infinite, what does that do to the constructs of this Infinite, that is a structure which is the evident visible truth of life? Both factions are therefore a dichotomy, because they have not evidenced the fact that the fourth dimension is a valid and concise statement of intelligence in which mankind is alive and able to see with a far greater clarity than with his physical senses. So how does he arrive at that development?

"Physicists do not believe that they can explain the fourth and higher dimensions or the abstract world that does not give its evidence through the measuring devices of the sense instruments of third dimensional physics. Then how will they be able to understand it and believe in its existence and be capable of measuring it? Now measuring does not mean measuring by the rule - feet, yards or miles, or in light years. There is an interdimensional measurement and there is a need to recognize the movement of the great interdimensional cosmos.

"Hence it is, the many physicists who are of great renown, such as Louis de Broglie, Max Planck, Niels Bohr, Einstein, Heisenberg, Schrodinger, Helmholtz, Eddington, Pauli, Gauss, and Dirac were mystics in many respects and believed in the continuity of life but were limited by the cosmology of the physical dimension in their attempt to measure the interdimensional cosmos. They were only capable of determining one factor of it, that factor that is within the limitation of third dimensional measuring devices! The physical factor would always be a limitation! It would be an indeterminate fraction of the Infinite.

"They did not think that it would be possible to ever develop a physics that would explain the non-atomic or fourth dimensional existence of life, because of the very anomaly that the third dimension is a fraction of the fourth dimension, and one cannot measure the whole with fractions. However, as we have indicated previously, the learning of the interdimensional cosmos takes place in the first building blocks of evolutionary life. The third dimensional foundation is the primary basis through which its extension into the fourth dimension takes place. This is the extension of the expanding universe as it is being experienced from its compacted nuclear form in its third dimensional equivalent.

"Unarius is an acronym for a universal articulate interdimensional understanding of science. Unarius, to this extent, contains the proposition, that the physics, if you will, of the fourth dimension, the interdimensional physics of the infinite, can be developed, and a mathematical system provided, to explain and describe the holistic nature of life while still living in the third dimension. Now this may itself seem a great enigma.

"How can the fourth dimension possibly be measured in the third dimension? It cannot! Not, of course, until the true measuring devices, which are man's mind, is moved out of the small room where the outer limits of the larger house can be experienced! However, that one room was still a part of the larger home or house or dimension, a spectrum of a dimensional factor of the greater Infinite.

"The interdimensional nature of life is partially viewed through the measuring techniques of electronic instruments, such as oscilloscopes, particle accelerators, telescopes and microscopes. These devices are definitely an extension of the naked eye. But it is the mind that is at all times developing the basic properties, which reveal to the inquirer who is stationed in a physical anatomy, the expandable nature of life and the coherent factor of his reality. That development is taking place evolutionary-wise in a dual manner. Certain scientists who have been mentioned, have previously recognized the duality of consciousness by having experienced the unified field of the infinite while living on an astral or higher frequency world, which is an evolutionary development of Homo sapiens.

Their scientific endeavors while on earth were to extend these fourth dimensional factors, of which they were already aware and had previously experienced as the reality of life's existence. To the extent that they were successful, they advanced third dimensional science. Hence, the very proposition of the expansion of life's perspective lies within the visible evidence of the nuclear forms of life, the nuclear forms which are themselves the cells of the Infinite. It is these cells which are the sparks of advanced scientist, which have evolved and whose eyesight has expanded. That eyesight is of course their mind, because the mind is a dynamic movement of energy that has in its resolution, developed a clearness, a lens ground to a greater perfection so that more light can be received on the film of their consciousness. More light means more intelligence, and more intelligence means that higher bands of energy of the Infinite have been perceived by that individual.

"Therefore, Planck and Einstein, Schrodinger, Bohr, Heisenberg, Pauli, de Broglie, and Dirac are such individuals whom we list as Unariuns. Why are they Unariuns? Because they contain in their own evolutionary development, the infinite concept of life. They were workers for the Light! Their purpose was to light up the interdimensional movement of Life! They were on the earth plane for this purpose, to advance the cause of humanity. They were using the logic and reason of the measuring tools of third dimensional physics, but at the same time, advancing the understanding of the continuity of consciousness of man's mind. They did not at any time, allow themselves to become soft or fuzzy in the application of the limited scientific tools that were being used at that time, because one must remember that they were pioneers on the frontier lines of the third dimension.

"In their respective ways they were explaining the initial nature of fourth dimensional physics, and it is still relatively recent that they have brought these new concepts into the realm of science. As they were all physicists, they advanced the understanding of the interdimensional meaning of life. They have done so within the measure of logic and reason, with the demanding analysis which is a product of such individuals who know from their own past life experiences of the reality of the Infinite. Yet they were faced with the problems of proving this reality through the limiting third dimensional measuring tools!

"In every respect, however, these physicists and others like them, are the very individuals who have been responsible for changing these mathematical devices for measuring light. In the work of Heisenberg, the very concept of the indeterminate factor of localizing the position of an electron, is an indication of his understanding of the reversal of energy from its positive to its negative polarity. Whenever a picture is taken of this tremendous movement of energy called an electron or a positron, the locality of its movement is polarized by the acting focus of the measuring instrument of the researcher - his own consciousness!

"Therefore, the full understanding of the product of the positron and the electron can never be concisely determined. It is on this basis, therefore, that Heisenberg advanced the understanding of the indeterminate nature of the forces of energy, which has been a great addition to the work of future physicists in resolving this puzzle of quantum physics.

"Niels Bohr observed, that the history of science teaches us again and again how the extension of our knowledge may lead to the recognition of relations between formerly unconnected groups of phenomena, the harmonious synthesis of which demands a renewed revision of the presuppositions for the ambiguous application of even our most elementary concepts.

"Schrodinger added an additional factor to this understanding, which had been developed previously by Planck, known as Planck's Constant. Energy moves in quantums or units; therefore, the attempt to measure the portion of that movement can only be had with the realization that it is now here and it is now there. The particles of energy are on the left side, then on the right side, whether being examined in some cloud chamber or upon some oscilloscope. However, for physicists to have any understanding of energy, they must have already had a previous awareness, which is the integrated nature of their own software. In other words, their computer is that much larger in its ability to contain information on both sides of the equation of life from its third to its fourth, and from its fourth to its third dimensional equivalent.

"The fact of the matter is that the *third dimension is only a mirror image of the fourth dimension.* One has had the total awareness by living in the fourth or higher dimensions, where, whatever one is experiencing or whatever the function of his present position in the material life, it is complete and whole and contains no separation. Although this is evident, the individual may not yet know of the physics involved in this totality of perception or consciousness. That is to say that he cannot explain the psychodynamic principle so established within his own mentality. All organisms, no matter what microcosmic features are being reflected, be it a caterpillar or a butterfly, functions according to the design that has been programmed into it. That is the nature of that species.

"As man is a fourth dimensional species, he can function without a physical anatomy. When man takes on the features of his negative polarity, an atomic body, he is naturally experiencing life from the negative polarity in the great movement of energy that is life, but he does not give any attention to it. He simply experiences life as all people do, without giving much concern to the mechanics that make possible this experiential movement of life's libido and drive. However, to be intelligent is to recognize all factors involved in the dynamic oscillation of energy. Therefore, anyone living in a fourth dimensional, higher frequency world, before he returns to a physical life, must begin to establish that very factor of the dynamic oscillation of his mind. This psychodynamics is normal and natural to him while he is functioning within the dynamics of an energy field that is non-atomic or astral. This is the unified field of a fourth dimensional consciousness.

"It is the atomic equivalent through which he is given the opportunity to look at the fractional unit, which is the nature of a lower frequency. This provides him with certain evidence of its dynamic oscillation. Now this dynamic movement of life, as it is reduced in its frequency oscillation, and in other words, slowed down, gives the individual the ability to recognize the individual parts of the energy field, that is the third dimensional, atomic pattern of life. It is a slowing down of the frequency structure of the Infinite within one of its spectrums of a certain dynamic oscillation. It is by means of this slowed down picture of the isotope that one can peer into the atom, and at the same time, peer into the atomic constituency of the physical anatomy to see the dynamics of physical life, thus gaining a greater degree of understanding of the interchange of energy.

"Such interchange resolves certain specific patterns of the movement of energy, atomic-wise, and takes on another resolution, for certain other features of life. To this extent, logic and reason is firmed up, because no positive progressive development can take place unless there is knowledge of the ingredients from which life is presently relating its picture. Therefore, it is in this respect that the interconnectedness between the third and fourth dimension is at all times bridged by the very individuals who have successfully strengthened their understanding of their fourth dimensional natures or their fourth dimensional energy system. That, of course, is their mind, the successful understanding of which, in this dynamic process, becomes a psychodynamic process, because man is the determinant factor, the consciousness of this great movement of infinite intelligence.

"The dynamic oscillation process taking place at all times, adds to the personal factor, a personal dynamics that is the body of man. The body of man is the energy that he has recreated and has become sensitive to in its recreation, as with a lens. Hence, an individual becomes more capable of seeing the relationship of not only the outer field, that he witnesses as a projection of his own inner field, but he sees the interrelationship of the two fields, the inner and the outer!

"Thus, man is the great researcher and is the product of the *Grand Design of Life*! In this distinction, man can be considered as a grand design of infinite intelligence!

"So we can say again, that the grand designer, infinite intelligence, has created the grand design of Itself within each individual! His physical anatomy is the secondary body and only a sub-dominant or demodulated factor of his primary body, the schematic of which is the map of a dynamic oscillation of the Infinite frequency patterns which is each individual's characteristic.

"The frequency pattern of each individual is therefore, the matrix that contains the understanding of the inter-dimensional cosmos. To the extent that this understanding is constantly evoluting, there are those who are just beginning with the realization of their own determinacy as a finite unit of the greater Infinite; as an expression of it. There are those individuals who have a greater qualitative development of this infinite constant, which they have evidenced from the polished lens of their higher consciousness.

"Such individuals such as Einstein, Plank and Bohr, have sharpened the lens of their minds to the extent that they have been able to add additional evidence only through the lower dimensional factors of man, his finite or third dimensional consciousness.

"The very fact that the acronym Unarius is printed on this page is evidence of its existence on another dimension, or else how can it have appeared except through the inter-dimensional movement of energy which is intelligence of the author? There must be an intelligence making possible the communication of this information from a higher transmitting station and also the fact that it could be received in a third dimensional equivalent to be printed in word form on the pages of this book. This is the evidence that there are advanced intelligent individuals who are capable of receiving information, validating the existence of greater intelligent minds, validating the reality of higher frequency dimensions of the mind!

"What we are leading up to, of course, is the basis of and validity for the existence of those advanced Intelligences beyond certain pioneer scientists who have added a new sum and substance to the understanding of physics. They have not yet been capable of rectifying into the dialogue of mathematical physics, dissertations within the community of scientists, the detection of the psycho-dynamics of life in a higher dimension, of which we are speaking. The higher dimensions have been classed under many names: mystical, metaphysical, spiritual. Although many noteworthy scientists have ascribed to such beliefs, they have not been able to add to the scientific nomenclature of their individual work - a mathematics for the movement of energy that is the oscillatory nature of the energy of this expanded field.

"Unarius is the expression and the proof of the existence of a physics which is an extension of third dimensional mathematical symbolism and the dynamic movement of the mind. The nature of the psychodynamics of the mind is a basic property and reference from which and to which all other factors of life are resolved. The aforementioned pioneering physicists were quite in accord, that the language of third dimensional physics was not capable of explaining the higher physics of the fourth dimension. Individually and as a group, they were serving to set up some foundations for preconditioning future dialogues in the latter 20th century, which would be the fulcrum for the *New Age of Logic and Reason.*

"Any information, therefore, that is capable of spanning the polarity nature of energy must be resolved from the unified field equation, which is lived in and by and with those individuals who have left behind the primitive lower fundamental concepts of life. They no longer work in theorems and hypotheses; at least to the extent of their knowledge of the mechanics of energy in dimensional equations of time and space.

"Fourth dimensional physics or interdimensional physics can only properly be explained by scientists who are relating this information from the latitude of their own interdimensional mind, and who can explain this physics without reference to the classical physics, which describes only half the picture, so to speak, of the whole. The very fact is that all of the aforementioned individuals are lending their own understandings at the present time from the fourth dimension in which they exist on a higher position, through means of electromagnetic lines of force, the highway of the interdimensional cosmos.

"That same highway is the individual's mind. The understanding, therefore, of the mind, which has been basically a progressive development of science in every respect, has been because Heisenberg, Schrodinger, Pauli, de Broglie, Einstein, Planck and Bohr, and many others too numerous to mention, added new perspectives of information that changed the entire picture of the atomic theory of energy.

"The fact is, they received this information through the highway of their mind that had been attuned to the scientists from a higher dimension. Therefore, the basis of all determinations will be the extent to which any individual advancing the boundary lines of his own mental equations can understand the dynamic process in which the mind functions. This new information will add to and will expand the previous limitations on which he was dependent from past-life understandings, which are obsolete in present relationships.

"To this extent, those individuals who are still hanging on to obsolete definitions, have been depending basically on determinations from some previous civilization, whether Greece, India, China, or Egypt, etc., and from which all people are presently receiving their reference points in the present, with the intent of understanding the finite dimension of life on their planet called Earth. The only means by which the future can ever be determined will be on the basis that it has no reference points necessary to certain definitions that had become crystallized and had been the basis of previous teachings. This is not to say that these teachings are null and void, but they did not at that time, take into consideration and apply the nature of understanding within the parameters of interdimensional physics.

"Science moves forward, at least from the third dimensional framework by extending the basis of propositions that had been advanced by Johannes Kepler, Rene Descartes, Copernicus and Galileo. They broke with an entrenched perception of life, and their scientific determinants were, in this respect, a constant breaking with past limited concepts as to the nature of mind and God.

"The future can only be known on the basis of information that is being related into the third dimension in the present age, by those who are futurists. Individuals such as Einstein, Bohr, and Planck were greater intellects than have been ascribed to them, as rendered by the work with which they were associated. They are interdimensional minds and contain a greater degree of understanding, now that they have been reinstated on the astral or fourth dimension. The infinite is ever infinite, ever expanding, ever continuous, and each individual is a dynamic oscillation of this infinite, moving forward in greater and greater understanding of the oscillatory principle of the interdimensional cosmos.

"The attempt to understand energy by breaking it apart into smaller and smaller quantums will not provide the answer, because the infinite lives within each microcosm of itself. Call it a quark, a lepton, or a muon, these particles are products of the infinite, functioning in a certain prescribed finite manner. It is necessarily a part of the larger infinite but contains its basic intelligence because of its present movement within its present frequency.

"Particle physicists are examining the descending feature of energy. Astrophysicists and cosmologists are examining the ascending feature. The proton is 2,000 times the mass of an electron, because it is the doorway or 'keystone' reflecting the pressure of the infinite cosmos, spiraling within its 'venturi tube'. All evidence of the life of an atom has so far indicated to the astrophysicists and atomic physicists, that the atomic spiral contains a compacted centripetal zone, and it contains a centrifugal zone and expanded rim. Neither the centripetal or the centrifugal movements are frozen but are constantly interacting between themselves. That interaction is the determination and indicates that the compacted mass, which is the nucleus of the vortex, is itself expanding in another direction and resolves into the centrifugal nature of the vortex in its sinusoidal movement. The interdimensional vortex is not singular. No energy form is singular!

"The vortex is the greater picture of the Infinite, moving in two directions from its own positive to negative position. As man exists on the physical, terrestrial planet, he is living directly in the middle of this dual expansion of the interdimensional vortex. Man is therefore viewing on his fulcrum, the microcosm through his particle accelerators, and he is viewing the macrocosm through his radio telescopes. But again, man is the instrument and is gaining in his research an understanding the dynamics and the schematics of his instrument. This schematic, his mind, is the lens with which he derives a conceptual framework of the contracting and expanding movements of the infinite. This is the present basis for research, using particle accelerators and quantum mechanics.

"Inner space and outer space can therefore be conceived to the extent to which the diagrammatic design of the infinite is revealed. We are presently revealing this Infinite to you by opening up the invisible field of man and reflecting in a diagramatic manner, the extent in which he is frozen in his inverted optical illusion of life. Scientists who have frozen their research within an optical illusion of life, within third dimensional boundaries, cannot rectify this optical illusion to see the image as it truly is - as a reflection of a unified field, contained within the isotopic expansion of a higher frequency of Infinity!

"The information then being related in this text is complete in itself, because it is related from the minds of scientists who, on the basis of their own mental development, are projecting this information into another mind. Such a mind has been stabilized and has developed a lens capable of demodulating higher frequency, energy structures into third dimensional, atomic forms, which become the word structure, the basic symbol with which man has learned to conceive the logic of his mind! .

"This process of interdimensional communication is the future for scientists. It is the future in respect that the properties of the third dimension are not themselves frozen in time and space, but are evoluting into great oscillatory movement of the Infinite. As one climbs a ladder, one climbs rung after rung and finally reaches a particular station or height from the ground from which he had initially taken the first step. The ladder is itself moving. Therefore as man moves, he is ever reaching a new level, a new platform. As man develops mentally, he is constantly moving into higher frequency structures and new relationships with others, but on higher positions in his perspective of life.

"A planet, as it is an energy structure of Infinite Intelligence, is moving in a sinusoidal or cycular manner. The oscillation of energy, which is the feature of life as an expression of Homo sapiens on planet Earth, is elliptical and expanding onto a higher circular orbit. Therefore, the ability for new information to be received into an instrument, is on the basis that such instrument must be constantly updated, renovated, and improved upon, whereby it is heterodyning between a higher carrier wave and its lower demodulated wave train structure.

"In this respect, man, the planet, and all other factors associated with a galaxy, are evoluting. The atomic constituents by which these mental constituents function are themselves changing. The atom itself, as a product of the fourth dimensional centrifuge is, in a sense, returning back to its causal basis, but always with a new and added intelligence. The isotopic nature of energy, which is the relativity of Homo sapiens consciousness, moves away from the lower atomic forms of life, such as gravity, momentum and inertia, to eventually live within a structure which is truly interdimensional, based upon the individual's ability to rectify the movement of this energy as a factor of his mind.

"Man lives in an energy body that is truly isotopic in its delineation. Therefore, the isotopic nature of the minds of those individuals who are responsible for the impressions that are the word structures of this book, are proving the interdimensional nature of life. Science will be the basic means through which the dynamics of this movement will be determined and for all people to be able to use as a learning device. Thus the highway between the third, fourth and higher dimensions will be well paved for mankind to be able to develop the essential ingredients of his own infinite capacity.

"Science is truly the highway itself upon which mankind moves, for even in its finite rendition, science is being changed into its more infinite rendition. All things are interdimensional! The very paving stones of this highway must be changed from the dense atomic form to the higher isotopic form. And the very ingredient that is now properly being reformed into the language of life and becoming a property of the scientific community, will become the pavement upon which man, in his lower development, still primitive or aboriginal in respect to their mind, will obtain a secure footing on this Highway.

"This Highway is therefore being paved on two sides you might say, and in two directions. On the fourth dimension, on the right side, the pavement is surfaced and strengthened purely from a nonatomic, crystalline material. On the third dimension, on the left side, the pavement is surfaced with atomic material, from information that has been derived from the opposite side of the Highway.

"The dynamic oscillation of this energy structure can hence be understood as a cyclic, dynamic duality. Each individual is involved in his recognition of the determining factors of himself as a product of this dynamic life process; the only means by which he will determine the real nature of consciousness. This is the established nature of the diagram that we have initially indicated to you as an interdimensional movement of energy, containing within its oscillatory intelligence certain developmental factors of life experiences gained by any Homo sapiens.

"Those who are called Unariuns are individuals who have advanced to the fourth dimensional perspective, who live within this expanded awareness, ever desiring to help man explain the infinite concept of cosmic creation. The information contained in this chapter is not a product of the physical mind, within the framework of a third dimensional proposition, but is a product of the interdimensional understanding of energy. That is to say, that the information is based upon dialogues, discourses, and dissertations from scientists who are presently living in a fourth dimensional energy structure.

"The isotopic nature of this consciousness is functioning on a higher carrier wave train and demodulated into the lower frequency of atomic language as a dissertation on interdimensional physics. The highway is, of course, the mind that is a development of the principles of interdimensional science, capable of discerning the lower mirror image of material or atomic life, to receive this electronic information as it is demodulated into the third dimensional, atomic symbolism of the word structure. We therefore see the great continuity between the third and fourth dimension, a continuity which is imperative to learn, as it is only in this manner that man wakes up to the nature of his own inherent infinite consciousness.

"Spirit therefore, is the all-inclusive, unified nature of consciousness and is a cell of the infinite creative intelligence, moving at all times throughout the dominion of its own consciousness. These are the beginnings of life for man, so to speak, in terms of his analytical understanding of this infinite proposition in lowest reference points on the third dimension; but at all times, oscillating with its own higher properties, his superconscious self, which is the cell that contains the unified, all-embracing, holistic understanding of life!"

Chapter 14

REGENERATION VERSUS REDUCTIONISM

"The literature of science does not contain basic information as to the regeneration of the basic amoeba, the basic bacteria. Instead the literature of all science, including the physical, biological and social sciences is illustrative of reductionism, using the technique of fractional research, delving deeper and deeper, into the substratum of the chemical elements that are the basic nomenclature of the Mendelian table of elements. Thus the whole history of science has been a table of contents based upon analytical factors compounding the fractional chemistry of life!

"This is all well and good with respects to understanding the regeneration of atoms into their molecular forms and into the cellular structures which provide for the natural habitat of life in the physical universe upon planets, asteroids and the satellites of planets, and indeed the sun bodies themselves. But it does not provide an understanding as to the initial nature of the regeneration of molecular forms. The desire of molecular biologists, physicists, medical practitioners, anthropologists and the paleoanthropologists is to understand the initial nature of life, by that they mean, of course, the evolutionary steps toward physical cellular life! And yet they are constrained by the lack of the understanding of the continuity principle, the space/time continuum from which stem the basic genetics of life. Hence, the methods of analysis used by these scientists are hampered by the limitations of physical consciousness which does limit any research involving the space-time continuum.

"You may very well ask why intelligent persons

equipped with the latest technological devices to peer into the heart of the atom, to peer into the living cell of an amoeba, or to peer into the cell of the growing fetus within the womb have not yet realized the interdimensional reality of Homo sapiens and planetary society?.

"There is of course a body of persons who have conquered the barrier of time and space; but have not been accepted in the community of scientists. They are the so-called mystics, metaphysicians, visionaries; they are the artists and poets, the writers of the comedy and the drama of man. But they are not welcomed within the narrow confines of research that is the privy of astronomers, particle physicists, molecular biologists and paleoanthropologists because they have gone beyond the constraints of the physical atom. They have seen the continuity of the micro-organism of third dimensional life, and they have had the gall to speak about life which has no basic derivatives within the physical framework of a physical universe. The third dimension is no longer the only world that is known by these individuals.

The metaphysicians who have an awareness of the fourth dimension as a mental mechanism, not supported by the physical body are therefore beyond time and space, are speaking another language, a language of another world. They have not included within their own discussions the understanding of the evolutionary movement of the physical atom from its fourth dimensional nature. They have not in this respect given an integrated understanding of the continuity of the electromagnetic flux fields of an interdimensional structure. They themselves have been speaking about the nature of man as he is evolving, is finally reaching a point where he no longer needs the support of the physical anatomy, but they have not given sufficient information as to how this development takes place.

"They have brought back a great deal of evidence, but evidence that is not supported in third dimensional physics, where the instruments of such physics would clearly show the continuity of man after his physical organism was loosened from his mind. Yet physical scientists demand a clear demonstration that the atom has had a beginning.

"What is the obstacle of individuals who demand proof of the nature of life through the examination of the ab-original past of man's remains under the earth of his planet; who examine the customs, folklores, and myths of peoples in different parts of his world; who examine plant, animal and sea life, to find some evidence which would provide a comparison and some validation as to the pre-history of Homo sapiens? As far back as one investigates into the physical past of man, he will not find the evidence for man's spiritual nature, because it will not be found within the genetic code of his physical anatomy.

"The genesis of man was not correctly indicated in the first book of the Bible. And yet within that statement can be seen a reference to the cycular nature of life. The six days are representative of the twenty-five thousand year period or cycle in which man has been functioning in his life to life gestation within the material universe. The regeneration of any physical object cannot take place without a substance that is the basic laser that makes possible regeneration of an energy that is of a higher dimensional material. That man was made from the mud of earth has a certain substantive basis when it is understood that man's physical anatomy is created from the sixteen atomic elements with the addition of the positive catalytic energy projected from the sun's corona. This action makes possible the regeneration of life through photosynthesis enabling man, animal and all other life forms that are basic to planetary life to evolve on terrestrial planets.

"This force has not been analyzed by physical scientists, as it is considered as a factor beyond their capability. Science is a body of information that has been so accumulated that it can become the genesis of new information which when added to the existing framework will provide for an expansion of the understanding of newer research under investigation. However such research must needs go be-yond the limitations of the five sense instrument of man!

"That is to say that the forces that are acting upon the planet earth are forces that are not only primary to the electromagnetic fields surrounding the mass of earth. There are other forces that act upon the earth plane from a distance. The entire galaxy is pressing down into the solar system in which the earth is but one of the nine planets! There are forces acting upon the planet earth associated with the other planets which are directly involved with the sun itself. Yet the sun is only one of countless sun bodies within the galaxy.

"In this respect, the understanding of life's activity as it functions on planet Earth cannot be understood by present information that has been accumulated within the electromagnetic field of the planet. The understanding of life within the entire galaxy, cannot be understood until there is greater knowledge of the primary field of the atom as the basic foundation for all material existence. The atom must have been generated by some greater force to make it active. It is in this respect that scientists have not expanded the nature of their research to add new information about the relationship of the atom within the fourth dimensional vortex.

"Metaphysics, properly understood is the knowledge of a higher physics beyond the third dimension. But, to be properly understood and to be conceived, it must explain the continuity of all life forces as they are propelled into their nuclear form, to regenerate the nature of their own material organism. One must lift one's eyes to look over the wall, to recognize that they have been imprisoned within the cave of reflected images which they have assumed to be real.

"Regeneration of knowledge cannot take place, cannot evolve within the mind of one individual until that individual recognizes his own evolutionary development, and realizes the continuity of himself as an atom. Thus within the particular discipline in which he is functioning he can extrapolate from the basis of his own experiences as he is a living proof of the evolutionary regenerative nature of Spirit, because he is a product of both the third and fourth dimension.

"There is no one who can remove himself from the duality of his own nature. When an individual who is concerned with the understanding of life involves himself in the sciences, but removes himself from the research data that he is collating, he separates the evidence of the evolutionary regenerative nature of himself, in-spite of the fact that he is attempting to understand the neurological basis of his brain structure, and the sociological and psychological implications of negative factors that are inflicting their damage in this community. He is himself, a part of the research in which he is attempting to resolve the enigma of a particular research project.

"In this respect, the molecular biologist who is looking deeply within the molecules maps making up the cellular tissue of the physical anatomy has separated himself from the physical organism. He has become immunized from the life force. It is in that respect that we speak about the vacuum that is the basic nature of present day science. The scientists in the large part have left out, in their analysis the evidence that there is no beginning to physical life! And if such beginning should be considered to be the birth of a child and through the developmental regeneration of the cell structures within the womb of a female, and if that is considered to be a beginning, then is that not an evidence of great intelligence? Because even the most immature scientist knows that the energy must have a basis upon which , and from which it generates its life force!

"Intelligence is the movement of a force which demonstrates its life in the form which it takes. Therefore, even the most constrained individual would ask the question, how does this regeneration of a single cell take place? Where did the force come from? Where did the intelligence come from? If the entire generic intelligence of a child is within the individual cell, how did it arrive there?

"In any discussions, therefore, of the regeneration and evolution of life forms, there is this basic question and the answer will always be distant from the questioner until he crosses that barrier line and realizes that the third dimensional atomic life is only capable of being regenerated when it is understood that its complete support system is on a higher dimensional energy field!

"Regeneration means that there is a force acting upon another force. The catalyst in the merging of these two forces results in the regeneration of new structures, which contain the intelligence to regenerate in another direction. The truth is staring all people directly in their faces. The basic proof is the corpse that was man! The corpse that contains the basic atomic structure of a Homo sapiens contains all the tissue was an instrument of complex intelligence. However that form by itself now contains no life! That is to say, it is supine, it cannot move, it does not have any support from a catalytic agent making possible its movement. The entire mechanism which is the complex nature of this anatomy is unmovable because something is lacking. That missing force is that agent that made possible the regeneration of the entire complex cellular function of this anatomy.

"This corpse represents, therefore, the basic mystery of life. Hence, we see that the catalytic force, which is in every respect similar to the radiation of the sun body, is of a higher polarized material. It is a higher energy.

"The corpse represents the lower frequency structure, and it is basically an instrument, inactive until an operator activates it. A material or mechanical instrument will remain inert and incapable of regenerating any information, except at the command of an intelligent force of the powers of ten. In other words, the body of man is a regenerated evidence of a higher intelligence, which demodulated into a lower frequency to become operable through the intelligence of the operator. But the operator, a function of another order contains the product which is the sum and total of a higher energy force field. Hence, the corpse is indicative of the lower nature or the lower structure of the field called physical life.

"The power that is responsible for movement of the atomic body is the mental body, a structure of the fourth dimension. It never had its development in the third dimension. It never had its development as an organism of atomic elements. Therefore the fourth dimension in every respect is functioning to make active the generic nature of a third dimensional frequency force field.

"The mind as radiant energy interpenetrates within the structures of the atomic body and manipulates the body to gain in this research the nature of all factors of Spirit. This is why the statement was made that any individual who has so set himself up to discover the hidden secret of life in some particular discipline, whether it be in geology, anthropology or in particle physics, etc., cannot separate himself from his research, because he is in every respect researching into his own identity! What good would it do to analyze the nature of the DNA molecules and not associate the findings within the larger framework in which that individual researcher is involving himself!

"Human nature is a development that is akin to realizing that all people have the same basic design. In this respect, they have a certain familiarity and gain a certain degree of confidence with themselves. There is a great deal involved in the evaluation of self in discovering the individuality and the spirituality of self. Because when man realizes that he is a product that cannot function except in the interrelationship with other forms as himself, he becomes aware in this respect, over evolutionary stages of life that he is a part of a grand whole.

"A proper study of man, is man himself. That means that there can be no proper resolution to the answers to the hereditary nature of genetics without an understanding of the impaction of man as he has lived in similar life forms and in similar life environments in the hundreds of thousands of formative stages of life. In other words, man has lived in the theaters of earth worlds during thousands of incarnations. How else can he learn of the efficiency of his higher energy field structure of the mind? How else could man attain any intelligence if he does not take down from the greater infinite those evidences which are gained through his own individual involvement within the scheme of life? Therefore, he can subtract from the greater energy field force whatever he wishes; he can extract any environment which becomes his support and provides a certain familiarity for him. Earth life, in this respect, is one of those facets or environments of evolutionary life.

"Now this information is not new. It has been presented in many different manners throughout the breadth of the literature of science, of drama, poetry and in the theme of sculpture and painting. Yet here has been a restriction in advancing this understanding of the continuity of life, that the mind of man is his fourth dimensional statement containing the entire panoply of his developing spiritual nature indicating the design of an Infinite Creative Force. Yes, man himself, not the physical anatomy, is an inner corpse when his mind is removed from it. But Man, that we speak of, is the evolutionary regenerative Spirit! So the proper research of man is not necessarily of the physical anatomy.

"It is true that this physical anatomy that is supporting his research project in his third dimensional life is important, and must be maintained so that he can continue his research. The purpose of life, therefore, from the cradle to the grave is to gain knowledge of the physics of one's mind which will provide one with the larger framework to extend one's mental eyes further than the walls within which he lives and maintains his physical life.

"The infinite can be seen to be the gestating nature of all processes of life, which cannot be understood until the lens from which man views life is polished . This must be the crystal where he sees a regeneration of the life from which is the containment of his atomic structures to reveal to him, the interpenetrating force from which the atom has made its entry. The proton of all atoms contains the passageway and provides the understanding of how atoms receive their support from their fourth dimensional structures, truly the originating agency for the lower structures that become the evidence of the five sensing structure of the physical anatomy.

"This, therefore, indicates that there is no beginning to life in terms of its physical beginning, because life did not begin in the physical or atomic dimension. In this respect then, the pattern of research will take on a different perspective, scientists will look in another direction. They will look into the agency of the energy structure not for the purpose of demanding its beginning, in terms of its physical characteristics, but to understand the continuity of its movement from another perspective! That perspective, of course, is another dimensional statement which enlarges the understanding of life to man. But life cannot be understood until the barrier within man's mind concerning the reality of life is removed, because life is not of a third dimensional origin!

"The barrier was constructed from life to life, brick by brick in the belief in one life. In this respect then, there was the beginning of the barrier. The wall was being erected and contained other factors that became the constraints of individuals who took on the costume of a religionist, scientist or political activist, to protect one's self should he discover that there is a greater understanding to be gained in the living of life on and through an earth body.

"Progressive evolution cannot take place on any level of life until the individual so involved, realizes the alternate movement from his own past life experiences, which are now completely restricting him from seeing the infinite which is directly in front of him at all times.

"The wail of a new born child is direct proof of the continuity of life. The last breath of man as he leaves his mortal coil of flesh is a direct indication of the cyclic interplay and the continuity principle of energy. Because, you see, it is in this respect that there is no straight line relationship in the ordinary sense of the sine wave as it functions in the third dimension. It is in this manner that the true advance will be made in the understanding of the cycular nature of life; a force which is Infinite in its breadth, depth, and in its complete roundedness; because in this respect one can remove all of the barriers from the constrained nature of life lived, in the physical body. One can then open up the windows of his room and let the fresh breath of energy enter into himself. That energy is the wafting of the subtle breathing of a higher intelligent Force.

"The catalytic nature of evolutionary life is seen in the higher structures attained through evolution. The love projected from a higher intellect is polarized to contain the full nature of the catalytic intelligence which is regenerating the Infinite in its nuclear and higher dimensional structure. Whatever that can mean to any individual is of course only conceived when he removes the barriers from his mind related to the belief that he is only a physically oriented, atomic body, that should he lose his body at so-called death, he will lose consciousness of what he considers to be life.

"This is the basic aneurism that is the threat of any individual in making any advance in the learning of himself as he is a product of a Creative Force. This aneurism is a mental sclerosis which creates a blockage within the structure of his energy system that supports the physical body. That mind is his consciousness, the window through which he sees the field of the Infinite. It may be a circumscribed field, because his window is very small. It may be a peep hole, because he has not added to the structural format of his mind, these properties which are the necessary factors to open up the aperture of his window. It must be done through each individual's own effort, because each individual is truly a scientist functioning to build an energy structure that becomes in every respect, a facsimile of the larger organ of the Infinite. These dimensional structures provide the ladder upon which man does regenerate the physics of his electromagnetic body.

"The electromagnetic body is the psychic anatomy, man's computer and is the proper study of man. It is this study that has alluded physicists and other scientists because they have given too much attention to the atomic genesis of life. Even though they have discovered the electromagnetic nature of the field that surrounds all organic and inorganic matter, they have not yet given sufficient attention to it, to map it's complex design.

"Man cannot die because he was never the physical organism. He simply inhabited it, contained it and directed it. Yet he can subvert his evolution if he does not understand the dominant force of which he has been created! The beginning of life in the third dimension is an interdimensional statement. The child growing within the womb has the need to develop a physical anatomy that could within the chamber of the womb gain the necessary strength to meld the mind of this individual with his newly formed atomic anatomy. The birth of the physical body integrated to become Homo sapien is a continuity from another dimension in which the mind of that child or that 'to be adult' was already integrated within the equivalents of a fourth dimensional system. The necessity of a womb is to provide for the structural atomic regeneration of the physical anatomy.

"Yes, that individual, that mind, that intelligence had already attained its complete development to the extent of its present evolution and is evolving again, but on the alternate polarity of the expanding and contracting plasma that is Infinite Intelligence! Evolution, therefore, takes place on two equidistant dimensions. Evolution cannot be understood by any scientist until he appreciates that it is a relationship between the physical anatomy and the mental anatomy. It is the mental anatomy that is now the proper study of man, and is the purpose of evolution. It always was, and will always be! In this time of the renaissance of spirit, man will truly awaken to the infinite nature of himself as he learns of his psychic anatomy. He will learn of the truth of his subconscious and his superconscious as circuits that are developments through many thousands of millions of experiences before the dawn of written history.

"Every book that has ever been written has been a basic tribute to the diary of man. Individually and collectively, the truth of life as a regenerative evolutionary development can be seen throughout the breadth and depth of present society, because it is the evolutionary design and the cyclic nature of life. The past ever lives within the present consciousness of mankind. If the past has been rectified the individual is evolving in a forward and progressive manner, his window is opened to the truth of the continuity of his own consciousness, of life after the physical death of the body.

Chapter 15

VORTEXAL MECHANICS AND THE VENTURI TUBE

"Philosophers and scientists have found it impossible to rectify the enigma of life as it is lived on terrestrial planets without an inward knowing of the macrocosmic and the microcosmic stream of consciousness that provides visible evidence that there is an anatomy transcending the immediate physical reality of an individual.

"The answers to the enigma of consciousness are expressed in the dichotomy between Aristotle and Plato, with Aristotle expressing the 'nature of reality' as the measurement of atomic life extensions, resolving into a system of classes. Plato has taken an opposite view of life and expressed 'reality' as a non-physical factor of life, and that the experiential nature of each individual's life evolutionary-wise, provides the individual with an understanding of the abstract nature of the infinite and of his place in the scheme of Universal Intelligence!

"We are faced here with the basic anachronism of the subject and the object, where Aristotle defends the position of the objective nature of atomic life and Plato, the position of the subjective nature of consciousness. Aristotle's logic and reason of life's appearancy follows from the statement that the objective nature of life is the apparent visible environment that can be seen, touched, tasted, smelled, and heard. Plato's reality was that mind was able to entertain its existence after the death of the physical anatomy, since the atomic environment was only one indice or polarity of the reality of consciousness. This has been the very same anachronism, which has been expressed in debates between philosophers and scientists since before and after the time of Plato and Aristotle.

"Indeed scientists can be philosophers, as philosophers in many respects, have been scientists. But there is a third philosophical equation that is a middle ground between the philosophy of Plato and the physical science of Aristotle, individuals who have abstracted from the Platonic ideals and have objectified the *'reality'* of life based upon their inner life, or the nature of thought, as a relationship between the atomic and nonatomic forms of life. These individuals have, in many respects, surrounded themselves within a conceptual understanding of metaphysics that disbands the conventional methods of research for the basic causal factors explaining the phenomena of third dimensional life.

"Metaphysicians postulate a reality above the basic physical reality of an atomic dimension - a reality that cannot be measured at the present time within the limitations of available instrumentation as measuring devices. Man's need to know of himself as a part of the universe has been a traveling dialogue throughout the past 2,500 years, and even long before that time.

"The founders of modern science, Copernicus, Kepler, and Newton, were themselves dedicated to this interdimensional awareness, and were aware of the reality of a fourth dimensional world beyond the visible evidences of the five senses. It was within this higher dimensional consideration that they conceived the principles of the finite appearance of the Infinite. In other words, they were able, in every respect, to distill within the finite perspective of a terrestrial world, the consideration and understanding of extraterrestrial principles to explain the movement of the energy. This movement of energy was made realizable to them because they were working on an interdimensional relationship of the objective reality of the atom and the subjective reality of the mind!

"The very statements made by these very aforementioned scientists, mathematicians and physicists have validated the dichotomy of consciousness which has been the basis for previous theories and hypotheses of life's reality, as with Plato and Aristotle. This enigma, of the subject and object, which is the parameter of consciousness, is a product of the researcher's mind.

"Looking into the deep space of the atom or into the deep space of the universe is, in every respect, looking into man's mind. This understanding places man directly within the moving nature of his own psychodynamics, the sum and total of an immense oscillatory, circular movement of energy, a sinusoidal matrix of electromagnetic energy wave forms, moving in a clockwise direction within an ascending and descending vortex. All factors of this electromagnetic energy field take on additional movements within the greater frame of reference of this tremendous plasma field, which is the indicative factor and proof of a greater Intelligence.

"So it can be readily seen that the subject who, at the present time is applying various measurements of an energy field, cannot determine the extent to which he, the researcher, a scientist or philosopher, is viewing this energy field as a reflection of himself. The interrelationship of the dynamic oscillation of his mind, as an energy field, is engaging with the dynamic oscillation of another energy field which form a duality, carrying the greater form of the creative infinite intelligence.

"To the extent that such a researcher recognizes the

polarity relationship of energy, he has been capable of determining the nature of his own individual relationship within the energy field, which he is attempting to investigate, and will result in a steady breakthrough in to new understandings and realizations of the nature of human consciousness.

"Certain individuals have an awareness that they are a negative polarity field equation and a positive polarity field equation within an interdimensional energy field equation of infinite Intelligence. They have arrived at a particular state in their evolutionary development, wherein they are capable of joining the material or physical nature of reality, their negative polarity field, with the non-physical, non-atomic nature of reality, their positive polarity field. This state of consciousness is the unified energy field interpenetrating within their own dynamic, oscillating, electronic mechanism.

"The basic perplexity and basic puzzle of all individuals, no matter what they call themselves, is their inability to see the relationship within the whole field. This enigma is indicated in any scientific research, in that there are fractional relationships involved in all research. This is true, because the unified field itself has been separated in the act of research. The interdimensional oscillating movement of energy, a whole number, has been separated by the researcher into an integer, containing another whole number.

"When this energy system, Homo sapiens, is not aware of the nature of his own integer, then he has fractionalized the infinite, at least to the extent of his limited understanding of the Infinite! With this fractional understanding of himself, he has developed a greater dichotomy as to the nature of man's purpose, and the determination of values in one's relationship to another human.

"Value systems of the individual proportionately become value systems of society, and the whole gamut of divisions and subdivisions take place within the entire nature of the social animal that is man, who is yet unaware of the wholeness of the integer, which is his spiritual self. Therefore, he has taken the finite picture of life within the framework of his physical senses and has fractionalized life, causing many perplexities, anachronisms and distortions, which again, as cobwebs, makes even more difficult the conceiving of an interior energy field system. This conceiving is, of course, the entire purpose in understanding life, because the interior energy field is the field from which all intelligence is attained; that is to say, as an interrelationship between the atomic and the nonatomic, the subject and object. This movement of the infinite within its associated respective polarities is the dynamic sum and substance of Consciousness!

"The entire nature and function of consciousness has been a dilemma amongst the purveyors of knowledge by certain individuals who have been considered to be intellectuals and had gained a certain degree of culture, based on the value systems of society. Inspirational factors of life make man more than the animal, make man more than the substance of his physical anatomy in its long, evolutionary history, from the primeval ooze of the initial bacterium.

"In this respect then, the present great leap forward in particle physics and in the adjoining field of cosmology, is a factor in overcoming the resistance that has been the basis of hard-core physical scientists, who with particle accelerators, have actually seen visible evidence of subatomic particles that contain specific intelligence in their movement.

"Particle physicists have seen the evidence of the subatomic streams of particles as a result of their collision in particle accelerators. This is evidence that activity is taking place beyond the evident physical senses of the researcher! The telltale evidence of the energy discharged as subatomic particles has been given the family name of leptons, mesons and bayons. Additionally, the mass and spin velocities of the particles have been measured.

"In this respect, it has been the belief of such particle physicists that they feel they are seeing the visible evidence of another dimension within the deep heart of the atom and seeing into inner space! On the other hand, there are selected scientists both in particle physics and other branches of science who are aware that a part of themselves is associated in their measurements in the determination and explanation of their protonic mass of the atom.

"It is true that a particle accelerator is the instrument that provides an image of such a discharge of electrons, positrons, and protons. But the determination as to the nature of this discharge as a path, observed in the cloud chamber or on monitors, is the determination of the physicist, who is in every respect, a part of the objective nature of his research. So how does one go about making a selective choice to validate the nature of energy forces, which are microcosmic substances of larger particles that make up the configuration of the atom?

"Science lays down certain rules in which the researcher must develop a consistent statement in the development and completion of his research project. The logic and reason for such analysis is well and good, and lays down various lines upon which sound research and its validation will be an accepted factor within the scientific community. But the missing element in research has been, and to a greater degree, still is, the hypothesis upon which such information is carried towards its final resolution, based upon the theoretical nature of such question being entertained.

"No matter then, what it is that is being observed in any dimension, one must always be aware that there is no possible means of maintaining objectivity within this measurement. That is, unless the individual realizes to a greater extent, that he can falsify, misrepresent, and distort the so-called objectives which he hopes will be the conclusion of his research paper; unless, of course, he takes into consideration the subjective nature of himself as a factor in the project being researched, namely the resolution of an enigma.

"In this respect, certain individuals of a higher development have realized how the subject and the object are part of the total field which is received on the retina of the eye. The subject has only one polarity, only one perspective of the whole field, and that polarity has to be understood as a fractional part of the whole; whatever determination is made from that fractional perspective would be the realization of the nature of but another fraction increment of the whole field. The fractional view can never be the complete whole; that is to say, a clear picture. It could never be the complete whole and carry the whole picture.

"However, metaphysicians have not as yet, themselves, determined to a greater extent, the physics of an interdimensional science as a recognizable science of this alternate nature of life. However, Emanuel Swedenborg, Immanuel Kant and the great Spanish philosopher Spinoza have described philosophically, a dialectic that contained an interdimensional physics of evolution.

"It is of no consequence then, to go around with the same little figures and using a variety of indices that are the crafting of the various techniques used in studying the motion of planets within the third dimensional framework in which these planets function, and attempt to ascertain answers to the puzzle of the creation of the universe, the creation of man, and the developing beginnings of Homo sapiens.

"In this respect, we then come into a basic enigma and a great dialogue that is presently causing a disturbance in the scientific community; the dialogue between individuals who do and do not believe that man is a spiritual creation. Those individuals who are themselves scientists, engaged in different fields of inquiry, in their attempt to join science with religion, have literally taken biblical statements from the Bible and abstracted them as a frame of reference in which they would resolve the enigmas in their particular scientific field, such as in archaeology, biology, anthropology, physics, and cosmology.

"They have attempted, and have, in many cases, successfully introduced legislation into various states within the United States, whereby the teachings of their perspective of life could become part of the curriculum, contrasted with the established teachings of the theory of evolution that has been an accepted feature since the past sixty years, based upon the evolutionary theory of Darwin.

"Creationism and Darwinism are, in this respect, an indication of the attempts by man to rectify this enigma as to the reality of life, establishing the reality of man within the broad perspective of the so-called truths that have emanated from the scientific traditions and of biblical teachings, within the frames of reference of physical life and spiritual life.

"These individuals who call themselves creationists, believe that the physical anatomy does not have a long history of evolution from the amoeba to the present time, of millions or hundreds of millions of years! Their ascribed belief is that man is a spiritual being and was never a physical being. It is the intent of these individuals to validate their teachings as a fundamental belief system, theologically based to validate the true spirit of Homo sapiens.

"In this respect then, Creationists are individuals who are attempting to validate this principle of Homo sapiens as a creation from an Infinite Being, to thus remove themselves from the perplexity of the dichotomy of life to which they are witness and of which they are a part. However, the Creationists have left out the vital factor of the interdimensional principles of energy, which contains the nature of infinite intelligence. That basic principle is the oscillatory movement of the energy itself, which is the foundation which inscribes man as a spiritual energy system. They have left out other factors in their declaration that man is a spiritual being!

"Yes, they have emphasized the spiritual factor man, but the spiritual factor of man in their determination is a flat-sided being that does not carry with it the understanding of its beingness. The science they supposedly use, based upon the history of scientific thought, is not a full science. That is to say, it is partial, because it does not recognize man's negative polarity in the evoluting of the beingness of man's spiritual nature!

"The spiritual nature of man is itself an evoluting, developing cell that contains the disparate nature of itself, within its positive to negative oscillatory field. Hence, these individuals have attempted to force upon the societal life of mankind, a doxology that is remiss in the logic and reason of the diagrammatic physics of the interdimensional science of life.

"On the other hand, we have scientists who ascribe to the evolutionary concept of life, ie., that man and all physical artifacts of the cosmos have gone through evolutionary periods of billions of years; that the universe is itself an evolutionary design. But on the other hand, they have not ascribed to, or accepted the spiritual origin of the atomic evidence of life! They are atomic scientists who, on the one hand, ascribe to the understanding that there is a continuity of life on a mental reference plane, but have no way of validating such hypothesis with the measuring devices of present instrumentation, whether in the discipline of physics or other disciplines of science. Therefore, they are left with the measurement of half of the polarity field that is the whole nature of life.

"The great difficulty with individuals who call themselves scientists is in their disbelief of the continuity of life after the physical anatomy has been returned to its base plane atomic constituency. As they do not have the ability to measure the full nature of the energy field of man, they are stopped in their tracks and are not capable of entertaining the possibility, or even the probability that the discipline of an interdimensional physics can be conceived within man's mentality while still living in a third dimensional energy field!

"This has been the basis of many scientists who, in many respects, have lived two lives. On the one hand, they believed in the accepted techniques used to measure the mass of an atomic body. But on the other hand, they were not capable of separating themselves and accepting the factual evidence of an alternate state of Consciousness!

"During the past thirty-five years the understanding of an interdimensional physics has been developed resolving the dichotomy of atomic physics and metaphysics. These principles contain a higher and more advanced understanding of science as a continuity of third dimensional physics.

"As such its basic proposition is a universal articulate interdimensional understanding of science, rather than the products of third dimensional mathematics and the attendant physics with which scientists wrestle with the enigmas of the atomic picture that is displayed on the screen of the monitor.

"We must remember that life is always ongoing! Life is the product of a tremendous movement of energy, a complex field that could shatter the swathed attitude of certain diehard individuals were they to glimpse only a fraction of it! This tremendous movement of energy is the nature of *Reality or Infinite Intelligence,* which can be viewed only partially at any one time in the evolution of any one individual. Scientists who are capable of providing a diagrammatical means to elicit a picture of this complex cosmos would be master scientists in their own right, capable of diagraming the physics of this complex electromagnetic flux field of energy!

"Advanced scientists have lived amongst mankind and have attempted to irradiate this knowledge in different civilizations. Spinoza, the Spanish philosopher, was such a scientist. But because of the time in which he lived in the 17th century, he was not capable of diagraming the interdimensional science that he described. But Spinoza, Kepler, Newton and others, were in their respective situations, able to add, and provide a larger frame of reference of the movement of energy from its great centrifugal expansion to its centripetal demodulation as universes, galaxies, solar systems, planets, man and molecules!

"The first breakthrough was with Copernicus, who indicated that the physical senses were only one small factor in the resolution of the reality of the larger energy field, wherein planets move about their respective suns. Such movement does not agree with one's own physical senses. But on the other hand, agreeing with the physical senses is agreeing with only one small part of the polarity relationship that is man. Therefore, on that basis, the entire perspective of life would be ordained within a narrow framework, when man did ascribe from his physical senses that he was indeed the center of the solar system. In addition, theological pre-cepts provided the church systems with great powers, as they took advantage of this false sense structure, whose misrepresented interpretation led mankind to believe that the teachings of the church gave the magnanimous understanding of a God who was harnessed to their own limited understandings of life!

"The continuity of the Copernicus theory that planet Earth moved around its solar disc was proven by the mathematical science of Johannes Kepler, in his three principles of motion, and continued with Newton until the present time. Science now has advanced to the complex understanding of the nature of the atom, in atomic physics, and coupled with stellar physics to determine the nature and function of star systems. There is a beginning of a cosmological interpretation and a new development of principles to explain how galactic star systems function.

"In this respect, the continuity principle of life has always been an experiential awareness of mankind of being a part of a whole, as he had lived with this awareness in previous civilizations. It was in the products of previous civilizations that were the responsible factor for the growth of man in this progressive evolutionary development, as well as a devolution into a retrogressive development! It is of great importance that this understanding of the interdimensional nature of man be retrieved as he has lived in previous civilizations and on distant planets. Psychic amnesia is one of the factors for man's inability to see his relationship with the 'real' nature of his *psychic self.* This is an important factor in the evolutionary nature of life.

"The other factor is that the Infinite is constantly moving in an ascendent spiral. To say the least, there is great intelligence operating in this vortexal movement as mankind is being directed similar to ships at sea to new destinations.

"Pilot ships, in this respect, are master scientists who have helped dock man, evolutionary-wise, throughout past civilizations, so that positive steps may be taken within each and every cyclic engagement of the larger plasmic field of the Infinite. This is a central purpose of life to be recognized by man.

"These master scientists who have been these pilots, have been called *way showers.* They carry on their ships a great light that provides the Beacon for man's ship of state, so to speak. It is these individuals, as scientists, politicians, philosophers, educators and artists, who have directed and piloted the ship of state for man in various societies, nations and civilizations on various planets within the galactic systems.

"Each individual truly is functioning with ingredients that mandate an understanding of principles which function as viable factors in the great stream of the Infinite. This great stream of life is the movement of infinite intelligence within the present conscious mind of man. Since the time of a great geological destruction 12,000 years ago, the momentum of a gigantic reactionary movement of energy has interfered with the progressive development of mankind on Earth. It has caused a giant explosion - planet-wise! The evidence of this geological fracture of an entire sheath of planet Earth, was the destruction of a continental mass of land in the Atlantic Ocean.

"Atlantis maintained the last great scientific technology that had been developed by these *way showers,* master scientists who have lived among man and had lent their support, compassion and intelligence to the planets from which they originated in the star cluster of the Pleiades. Of course, the true spiritual nature of these individuals could only be recognized through their works and in the attendant nature of their involvement with the people whom they had helped to recognize the spiritual ingredients that lay within themselves.

"The present 20th century therefore, in this respect (as we are bridging a large series of events, from twelve or more thousand years ago up to the present time) is the reemergence of all factors of life that had been lived by the earthman, when they were citizens of Atlantis, as Atlanteans!

"The island continent of Atlantis, before it broke into three separate masses, contained contingent groups of people who are presently living today in the 20th century. Whether they are in the physical anatomy or in the astral anatomy makes no difference. The present development in information technology is the reestablishment of an interdimensional science that was taught and practiced in Atlantis, but was changed to a third dimensional science by those who would not advance in their own mental development! They were then detuned from the spiritual principles that had been introduced, and taught by master scientists, teachers and administrators of the Atlantis society!

"These individuals had become negatively biased as they did not concur with or add to the ever-moving mental nature of the interdimensional cosmos. As man is a proper part of the Infinite, he is the subject, and becomes the object of the larger field - the objective or positive field. Man must therefore always interrelate with the positive field so that his atomic negative field receives the positive input to gain access to even greater strength and power, containing the greater intelligence of a spiritual nature. This constant sinusoidal movement, always moving cyclically, takes on the positive to negative appearance. The positive polarity imparts its Force, its intelligence to the negative polarity and gives the negative polarity its positive contour in the next ascending spiral! Aborigines in previous civilizations have conceived the sine wave as a snake, for it is in this movement that an information is exchanged or interchanged between the coiled nature of itself.

"In this respect, the nature of life is a movement, not of the physical anatomy, but of the energy anatomy, the mind. Atlantis perished below the waters of the Atlantic Ocean because the negative polarity of man refused to move up into the next and higher quadrant, which would stabilize man on a new platform and on a new psychic development!

"Hence it is, that there is nothing new under the sun, because all that is presently a factor of life has already taken place. But there can be something new! In fact, it must be so - a new perspective, a new resolution, an impression of

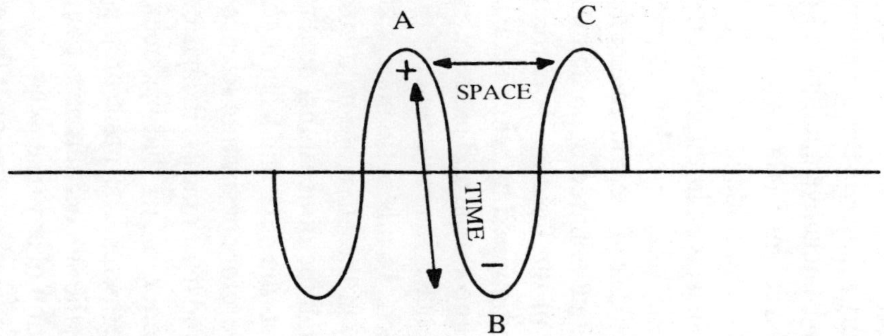

SPACE FACTOR - A to C

TIME FACTOR - A to B

life that reflects a higher consciousness, a higher latitude, and a greater understanding of the surrounding energy field! This therefore provides new information of the realization of the immortality of man, the nature of man as spirit, which is the whole number of the Infinite.

"Twentieth century science is the development of, and reengagement with, past negative applications of energy. It is also adding logic and reason to the awareness of the negative or half life, or the half understanding of the reality of man's inner being. Unarius is, in this respect, a school that is based upon the reality of an interdimensional cosmos. It is a school that is not made of the concrete of an atomic element. It is a school that is based upon and built from the nonatomic elements of the Infinite. It is created from infinite intelligence and applied by scientists who are interdimensional in their knowledge and usage of energy. It is a Science of Life that teaches the basic concept that man is the subject and the object. In this basic understanding of the physics of the energy in the alternate nature of itself, man learns to move through the dimensional quadrants that are his life from one life to the next, and from life to life.

"The principle therefore of any true science of cause and effect, of frequency and harmonics, is the basic statement of interdimensional science, a fact that has been known and practiced by all wise men. The basic proposition of any science is that man is the subject and object and is the chief factor in the proposition of any research.

"The study of man as one polarity of the diagrammatical movement of the energy forces of the infinite, provides him with the ability to abstract from the constant movement of the sea of energy, and to develop a perspective which will hold him and give him strength to set his course in a direction which will add a greater knowledge to himself. His ship will be laden with the gold that can be mined in this higher dimension; an element which will not rot or rust, but which will provide him with a higher foundation to make his voyage into the interdimensional cosmos, and set his sails to arrive at the Port. Here, be met by those very individuals who had laid down the guidelines in previous civilizations, in the lifetimes in which he was learning of his inner self!

"He will meet illustrious personages: Anaxagoras, Plato, and Socrates, Akhenaton, Hermes, Einstein, Cellini, Leonardo da Vinci and other personages who have lent themselves, mentally-wise, to mark a direction in the worlds of man, to take and lead them to the inner portals of their own mind, whereby they could open up their map of life. Here they will meet with new friends and teachers who had helped them on Earth, to give them the strength to continue through the raging seas of the world of the atom.

"Only with knowledge and strength of wisdom can man learn of the reality that life presents to his mind. Such a mind, having been so polished that it reveals the incredible picture of beauty, of balance, of calm and repose, must be within each individual before he can view such a world. Therefore, we say to you as a summary, that to be able to objectively view the real worlds in which man lives in repose and in great contentment, in tune with the Infinite, one must attain to that state of mind, because the subject can only reflect the object and the object will reflect whatever is the subject of itself!

"The recourse is that each and every individual has an energy system which has been created by Infinite Intelligence, whereby he can refract this knowledge in which his instrument will reveal to him the nature of its refraction. It will reveal to him whatever he has done to his mind! He will see either a world of ignominious peoples, living in mental duress as pictures of great distortion, completely unaware of the greater reality, or he will see people who are so integrated and balanced that they project the beauty of their own inner nature. This inner nature is shown objectively as a scintillating movement of rainbow colored skies, and where the cities and homes are radiating the many-colored hues of energy which reveal the evidence of a higher intelligence, permeating through every home and every individual whose beingness is revealed in this auric splendor.

"Science is a road map in all respects,to help man to rectify the energy which is his sensing device, to provide him the opportunity to see the real worlds in the fourth and higher dimensions. In this greater desire, there is a greater motivation to address oneself to the basic proposition of life to refine and sensitize the instrument of his mind.

Chapter 16

THE NON-PHYSICAL NATURE OF MAN

"The anachronism of all descriptive literature attempting to describe the non-physical element of man, but more properly, the science of the fourth dimension, as the new science of quantum mechanics, makes no mention of the schematic of the electronic energy field supporting the physical atomic body.

"It is an anachronism that man does not understand the complex diagrammatic energy system that is his electronic body. It is indicative that there has not been, up to this time, within the past 2,000 years, a mind capable of laying down a map, illustrating the physics of this electronic body.

"It is our purpose here to lay down a map to describe the closed system of the electronic transmitter from which all physical atomic life factors are derived, as well as evidence of non-physical factors that provide man with a recognition of the continuity of his life after death. Even the aborigine was aware that there are intelligent people on other planets, as they viewed the starry skies filled with the lights of different star systems, indicative of the continuity of all matters and forms of living, breathing organisms that are manifestations of a functioning intelligence of the First Order.

"The latter part of this 20th century is known as the Age of Space, and of Computers, the Age of Astrophysics and of Particle Physics. This is therefore an age for the resolution and recognition of this great matrix of energy that is the primary organism for the life of terrestrial man, animal, plant, vegetable and mineral.

"Scientists presently functioning from a fourth dimension, higher up in the spiral vortex, have projected a diagram containing the internal map of man, as well as the external map of the universe. This electronic schematic can help to create an advanced technological civilization, aiding the understanding of the movement of consciousness proceeding from two parallel dimensions of atomic and nonatomic forms.

"The 20th century contains the prehistory of past civilizations, and the evidence of advanced technology, initiated by scientists in previous civilizations; some known and others unknown. Unknown civilizations buried within the depths of the Atlantic, Pacific, Indian and Arctic oceans, and under land masses of earth are being revealed indicating the technology of these ancient cultures.

"Mankind has existed previously with knowledge of the interdimensional nature of consciousness, that there has been an abortive attempt to successfully eliminate from the present memory of man, the recognition of his past. But the fourth dimensional nature of man's mind is such that it is constantly pressing down from its higher station, from a higher frequency dimension.

"This is the cycle, now in the latter third quarter of the 20th century for man to awaken to his past, when he lived more closely in tune with his spiritual self and with spiritually advanced peoples who exist in higher civilizations. In what other way can there be proof of the existence of the unified field equation than with a unified consciousness?

"Man has tasted this proof of a higher intelligence! Advanced Homo sapiens have designed, erected, and developed civilizations that functioned in a well-organized and balanced manner, in which man was being taught the basic physics of his inner nature, applying this physics to the outer environment by crafting upon it the map, revealing the nature of the universe.

"The present 20th century is the sum and total of the continuity of all past lifetimes. It is an information network that lives within all humanity! It is this information that accesses man to past civilizations upon this and other planets in the Milky Way Galaxy and other galaxies. Such communications are based upon the knowledge of an extended physics that enabled intelligent Homo sapiens to develop technologically advanced civilizations.

"In order to gain access to this knowledge, one must remove the blocks that have been the rationale raised by individuals who did not desire the principle of the continuity of life to be expressed within the developmental nature of life. These were individuals who wished to hold on to the self-imagined power that they developed with respect to their superior knowledge of the primordial elements of life; individuals who found themselves in a position where they could receive the accords of people by accepting the worshipful attitudes of millions of their subjects.

"The aborting of the progressive development of man, created a block in his ability to attain to the realization of his interdimensional intelligence. The continuity of his consciousness has been shut down and is the basic reason why there is amnesia amongst mankind today, with respect to knowledge of his past. The libido and drive expressed into each individual as to his own relationship to the creative cosmos, does not ever stop in its expression of itself, to energize the lower nature of itself in his atomic constituent.

"In this respect, the proton can be considered to be an expression of a higher frequency dimension, and the electron, an expression of a lower frequency dimension. But that electron must itself express itself in a manner and way, and in a direction which will re-express the nature of the intelligence that exists within the protonic factors of itself. Not to do so causes great disarray and a breakdown in the movement and expression of the intelligence that is being resolved into these equivalents of energy.

"Man's higher nature is a protonic electronic element of Infinite Intelligence. The electron is man's lower nature, (so to speak) as it is expressed in lower frequencies of the infinite in a terrestrial dimension, and loses its direction when it does not continuously engage consciously in the oscillation of this higher energy field. Such discordant intelligence expressed, is distended, and lacks the full nature of its primary motivating source.

"Therefore, as man lives in the physical, terrestrial worlds, he reflects the disorder, disarray and the disintegration in his mentality, caused by the separation of himself from the primary *Force Field.* Hence, all diseases are caused from a separation of this Power, this Force, this Intelligence, which is the necessary impelling Intelligence, charging the electron and man, and activating his intelligence.

"The atom is a television screen upon which images are reflected, to provide direction in the life of its energy quotient. In any description of the mind, one must recognize the features of this mind as it recreates itself through the atomic nature of its physical anatomy, powered at all times, of course, by its protonic Force. This protonic intelligence is expressing the essence of its higher nature to reactivate the lower electronic energy force in successive lifetimes in which the mind, the fourth dimensional energy structure, recreates an electronic body; that is to say, a body that is manifest in the third dimension as a five-sensing instrument.

"The alternate movement of the *Cosmic Force* from one lifetime to the next, with Homo sapiens is an expression of the descending spiral in the sinusoidal movement of energy. The rise and decline of civilizations is indicative of the cyclic movement of energy. But it also indicates the opportunity for the energy force, which is mankind, to ascend in this sinusoidal movement of the great cycular manifestation of Infinite Intelligence. In the history of man on earth the descending nature of civilizations has been revealed in Atlantis, Egypt, Greece and Rome, and other civilizations, and into the 17th, 18th, 19th, and 20th centuries, with the example of Portugal, Spain and Great Britain. In the present 20th century, the United States of America and the Soviet Union are the two expressive positive quotients in the ascending spiral of society; yet not altogether similar.

"In every respect, although we will not move too far away from our central discussion, the United States of America and the Soviet Union are also a basic proof of the principle of the sine wave, containing the positive and negative movement of life in the central arenas in which it is presently being reflected upon planet Earth. The forces that are posed within these two societies, are indicative of the continuity of the people from their past-life expressions, presently re-flected into the societies of these two nations. The United States of America contains in a greater manner, the positive bias for understanding the proposition of life from a higher frequency dimension, by reflecting within it, certain factors which are expressive of this knowledge.

"At the same time, it contains within it the negative polarity of its own self, because there is no equation of life that does not contain in its indices, the opposite nature of itself. In the Soviet Union, we see again the identical reflection of positive to negative; the contrasting expressions of life that are the factors of life in that nation, revealed as the lower end of the sine wave in its negative polarity. Certain factors within the nature of life in the Soviet Union are indicative of a vacuum with respect to the spiritual quotients, which are the dominant nature of man's mentality, the dominant nature which is his Superconscious Self, the fourth dimensional and spiritual equation of man.

"Therefore, in this respect we see the Infinite constantly revealing itself in all movements of life within man and animal, plant, and mineral; but in man, of course, this movement is much more complex and much more evolved. It contains, in this evolution, a greater indice of the higher fourth dimension, as it has been rectified for each individual in his continuous evoluting; his fourth dimensional life expression, and then his third dimensional life expression - a plus and a minus, positive to negative expression of the sine wave principle of life.

"Therefore, man has moved in this oscillating manner between the two polarities of infinite intelligence to the extent where he has polished the lens of his own refracting electromagnetic energy. This polarizing process becomes the intelligence of that individual who is able to see more clearly the nature of himself in the expression of his life in both an atomic and a non-atomic body.

"The mind is therefore truly a vast computer that contains hundreds of billions of trillions of bits of information, but so integrated that these bits are harmonically linked to each and the other as they are formed and reformed into different capacitors, which reflect the nature of their design; a body that contains the higher frequency code of dimensional expressions of the Infinite. This mental computer that is being expressed as the third dimensional computer, is in every way, the proof of its existence.

"Present computer science is based on knowledge that has been inputted into man thousands, even hundreds of thousands and millions of years ago, when certain responsible scientists and mathematicians initiated the development of computer science, in previous, high technology civilizations and had learned of the interior imprint of this great electromagnetic mechanism that is known as the mind.

"We see here in this respect, that the protonic Forces of the Infinite are constantly moving into man, expressing itself to a greater degree than it had for many thousands of years! And we say, thousands of years after the decline and the destruction of Atlantis, where computer science was the very visible factor of everyday life, used in such a way that there was no need for the vast array of instrumentation as is presently the scene.

"Certain advanced persons were capable of using their mental mechanics, their computer, to elicit the necessary results. They could in this manner, gain access to information presently on their computer tape in their computer library and apply it to technological works. Of course, this is the entire purpose and intent of computer technology, and it is being learned again by mankind.

"The information, therefore, that is necessary for man, for him to attain full access to his protonic nature, is the understanding of the circuitry in which this protonic force functions. In order to gain that understanding of the Superconscious Self, which is basically the proton of man in his expression in his present material life, he must know of the other two components or circuits of himself as well.

Knowing of one without the other would necessarily thwart him in his desire to gain an understanding of himself. To know the reality of himself, he must know the relationship of his subconscious to his mental self and the *Protonic Force* contained in the *superconscious* circuit, which is his fourth dimensional nature, his spiritual self!

"The vast proportion of mankind has functioned on the lower circuit, the subconscious, which is being powered by the mental and the superconscious self. But as we have seen, this power has been diverted, and a dam was erected. That dam was the impressiveness of the visible five senses, as it was capable of providing certain privileges and powers to anyone individual. The overemphasis given to the visible nature of life in the atomic world exaggerated the subconscious. That is to say, it gave a greater force to the movement of energy as it was functioning in its sinusoidal manner, and working as all information must do, on its positive to negative duality.

"The subconscious, in this respect, therefore builds a physical anatomy each lifetime, with a new physical body created through the womb. The individual matures and reorients himself or herself within the environment of his social system and obtains the sum and total of all necessary information to live within the physical anatomy. Such movement is obtained from the electronic body, associated with the circuitry of the subconscious part of his computer. Memories of all past lifetime experiences are accessed from the mental circuit. It is this tape that is functioning in a space-time-continuum, containing all indices of life experiences accumulated from all previous lifetimes! These experiences imprinted as electronic wave forms, regenerate a new tape, so to speak, a computer tape, with all information necessary to enable a new physical body to function. This new physical anatomy functions from information accessed from the subconscious circuit, coded for the expression of life on an atomic frequency. The molecular nature of each cell responds to this electronic code, from imprints of previous life experiences formatted in sinusoidal vortices.

"All experiences are accumulated through inner relationship with the third dimensional environment, re-accessed into the computer tape of the subconscious circuit, so that information of a physical nature is imprinted, thus regenerating similar energy information wave forms. In other words, these acts, harmonically re-engrave and add to that individuals present life, similar associated past experiences from life to life, to the present lifetime!

"Such an individual who is born into a terrestrial planet, who has been overly concerned for the material expression of life, has imprinted upon his subconscious circuit this concern from previous lifetimes. He is, in this respect, considered to be a materially-biased individual. That is to say, he lacks access to the *higher energy force fields* which contain information of the interdimensional nature of himself and of the existence of life in the nonatomic worlds. Even though he has lived in these worlds, his memory of them is not sharp. In many respects, he has *psychic amnesia*. You might call it spiritual amnesia, which is more correct.

"This individual is therefore inordinately concerned with advancing himself or herself within the expression of his society, to become well-known, to all intents and purposes, in any expression which he has previously developed. The desire to re-express the ego is the basic motivating factor in that individual's life, because this is what he has programmed into himself, life-after-life!

"Man is an electronic mechanism, and that electronic mechanism will only reflect what has been in-putted into it. Therefore, there are, in this manner, hundreds of millions of people whose only concern is to provide food for the stomach, a roof over their head and the delights of the five-sense body! Some individuals may not be satisfied with a five room house or even a ten room house, but are satisfied only with a mansion. The satisfaction is never complete, because of his misdirection of the energy which is that individual's protonic force. Therefore, with such individuals there is in this misdirection, disease, both mental and physical - a breakdown of purpose and intent in his life, which, of course, affects other individuals who are so associated harmonically with that individual from previous lifetimes.

"Previous lifetimes contain the *continuity principle of life*. It is the important and very necessary factor in the understanding of man and of his evolutionary prerogative to constantly add to the understanding of the interior map of his mind so he can set his sail, or using more modern terms, ship aboard his spaceship and journey to the outer planets, into other solar systems and galaxies. He will learn in this manner, of the great equation of the Infinite as it is demonstrated in the innumerable, countless planets of the universe. He will recognize the continuity of this greater picture of life as a reverse mirror image of what it contained in countless dimensional equivalents that are lying on a higher frequency relationship within the great expanding vortex of Infinite Intelligence.

"You say this is difficult to conceive, and that your mind is staggered by the immensity of the picture? And this is true, because until one has himself taken this knowledge, and it has become an understanding in the application of it in one's life, it cannot be a factor of himself and be his *wisdom.* The wisdom contained in the mind of the sages, so to speak, is the proof of their evolutionary development in which they have learned as students and as teacher. As student and teacher in each and every successive move up, within the evolutionary spiral of the Infinite, they have exchanged the basic relationship of themselves in learning of the higher principles of the continuity expression of life.

"Learning of the higher equations of life as advanced teachers, they are presently expressing life and expressing this knowledge to the lower spiral, into the terrestrial dimension. Man is at all times, expressing the sine wave principle of life in every movement of his mind. Therefore, it is quite obvious that as one learns of the eternal design of the mind, he is learning of the eternal nature of the Infinite, and he is also moving in-tune with the Infinite.

"In this manner does he gain what has been popularly *known as 'peace of mind'.* He gains integration of the sinusoidal movement within himself. He is integrated with whatever position with which he is presently involved in some body that has been fabricated from the energy system, the dimension in which his mind is being activated. Hence, there can be integration in the third dimension, where the mind and the body are operating in a balanced manner. Integration on the fourth dimension and on the fifth and sixth, etc., must be learned by each and every human being as he enters into a new stage in his evolutionary development, and is acquainted with new equations of life.

"We will not taint this image, this understanding, using the obsolete metaphysical explanations that have been developed in previous civilizations, and which are certain outcroppings carried within the memory of a variety of peoples who have not yet conceived the *interdimensional understanding of science* and its implication in the life of man. The importance of this knowledge is of the physics of energy. The movement of energy is, by its very nature, the expression of the constant balance that is reflected by Infinite Intelligence in its polarity exchange, as it is constantly charging and recharging Itself.

"The essence of this understanding is to realize that the most minute particle within the atomic mass of a terrestrial dimension is receiving the nature of this charge, and is discharging itself, either in a complete integrated fashion or in an incomplete fashion. If it is incomplete, the charge sets up a static field. Thus, it is incomplete in the sense that it is moving in a closed system and is not expressing the nature of itself in every respect, even in its most minute form. This could be considered to be a dam, a block, as the energy expressed not being able to move into its adjoining dimension is detoured in its journey in a vastly deformed manner. It is this deformity that is the result of all diseases, the result of limited information received from protonic force fields of Infinite Intelligence.

"Therefore, that energy expresses itself in a warped fashion, mutated in many respects, and malformed. This malformed, mutated, deformed energy re-expresses itself as it must, into the third dimension, in the form of some atomic form of life. You would therefore ask for proof. In every respect, the proof is an equation that can be recognized as obsessions.

"William James and Charles Pierce explained the logic and reason of energy expressed in a deformed manner as obsession. Obsessions are the malformed mutated expression of the primary reflection of the energy of the Infinite Intelligence. This deformity can be seen in the face and form of man; in the immediate change of any individual when he is discharging a static field of energy that is stemming from his subconscious circuit, containing some past-life trauma. Such trauma may be fear of ultimately losing the physical body, or phobias associated with the present negative events of society and the various interpersonal relationships of man to man.

"A house burning down, a great tidal wave, an earthquake, a volcano, a wound in the physical anatomy by a bayonet or revolver, an atomic explosion, etc., are pictures that are imbedded into the circuit of this computer tape, the subconscious. It is the mental consciousness that contains all the previous lifetime experiences re-expressed through the subconscious into the physical consciousness of man.

"Unpolarized negative energy is a mutation resulting from the very force of the trauma witnessed, sensed, and expressed by an individual in the moment when he was engulfed in the waters of an ocean, immersed into the hot lava of a volcano, and of other innumerable negative, life experiences. These experiences are immediately reconverted into electronic wave forms of a fourth dimensional frequency, and endlessly repeated in future lifetimes!.

"These negative life experiences are re-expressed again and again and again, in each and every successive lifetime, and in a regenerated manner. The mutation of energy proceeding from the Infinite Creative Intelligence is a vicious circle, which constantly envelops mankind, wherein he cannot emerge from the heaviness and the overdeveloped emotionalisms, so associated with these traumatic incidents. These traumas were not caused in this present lifetime! They were in every respect, an endless repetition of the first cause, which was a factor of that individual's experience in his relationship with society.

"The mutations of mankind are reflected in society as individuals who have been crippled by muscular dystrophy, multiple sclerosis, cancers and psychosis reflected in the mental behavior of mankind. In every case these are obsessions! These obsessions are reflected and expressed from an individual's own past lifetimes when they reengage and through frequency attunement to visibly change the expression of his life.

"We did indicate that the changes in the physical anatomy are caused by the associated mental and muscular diseases. Disease is a mutation, contagious to the extent that the mind functions as an interdimensional movement of energy.

"This means that individuals who are so-called dead, are living in a non-atomic body in a fourth dimensional world; and all persons who are living in a physical body are constantly receiving the nature, or imprint of their consciousness. More properly, all thought is being transmitted to other individuals. All life is a constant interchange of energy in accordance with the frequency and harmonics of the energy sine wave structure. Each and every individual on this and other planets throughout the universe are transmitting and receiving the sum and substance of life that has been lived by each and the other individual, according to the frequency they have inscribed, within their energy body.

"The fact is that all people are hooked up in this manner as receivers and senders. This understanding will be the greatest help to break down the barriers presently causing the deflection of an individual from the knowledge and reason for his present deformity.

"Not only does man recognize the impact of his own past, associated life experiences, which have become mutated, but he recognizes his past, associated life experiences associated with other people. Of course, since the negative intent of life has been highly developed and charged within the atomic form of life, and since it is the frequency in which all man is living, he is therefore open to the information contained on that frequency. Man is an open channel until he learns how to change the channel! Until one does, one is open to certain negative forces which are aligned harmoniously and automatically received into the mind. One is receiving such information through the highway of his subconscious circuit and re-expressed into his physical anatomy.

"All of the ills and ailments of mankind can therefore be seen to be a constant interrelationship, charged and discharged from oneself to others, and from others to oneself. This great interchange, where the knowledge of the electromagnetic spectrum, the knowledge of the movement of thought itself, can be the greatest factor for the health of the individual and society.

"The knowledge, therefore, of the mechanics of the mind is not for the sake of sitting in some quiet library room or in your home and pouring over the books and tracing the circuitry so shown on the pages of the book. But rather, in applying this knowledge to oneself first and foremost, by developing the expertise of this knowledge one does master the movement of one's thoughts.

"As all persons are involved in charging and discharging energy from their past, one learns how to discharge and rectify any negative energy wave forms so as to receive from a higher frequency band width. This higher frequency band width is formed from an intelligence projected from a higher spectrum in the infinite equation of life.

"Healing is therefore a knowledge of the electronic nature of thought within each and every individual. The Infinite is charging and discharging from and to itself, and the mutation that is an obsessional structure of energy, is the deformed nature of energy that is not being recharged with the intent of its true self from its higher polarity. Obsessions are the sum and total of the thought structure of any one individual, which contains a static field structure, carrying such information piggyback, so to speak, as out-of-phase sine waves.

"Think of these individuals, therefore, as energy constituents, rather than in a form of atomic structure. One can then be more capable of realizing that as he would look at a polygraph or at the energy wave forms stemming out on an oscilloscope, he can see the more complex energy waves in their harmonic association, integrated within each and the other, and forming a stronger and more modulated wave form. This is, in every respect, the signal of the sound and of the picture that are now so integrated that they conduct a base plane frequency of some information. That information is carried over the highway of each and every individual as his thoughts.

"Thought forms are basically the energy wave forms that have been discharged to the subconscious from the mental and Superconscious self, and are being directed into the prism structure of the physical anatomy. This atomic prism contains the necessary component parts to interrelate, integrate, and provide a complete overlay to become the imagery necessary to contain life on the atomic level, through a physical anatomy. Therefore, an individual can seem to become the visible evidence of the information, which is constantly stemming into him from another dimension.

"When an individual is inspired to express some factor of life, whether it is an invention or the manifestation of the life forces, or in sculpture, painting and writing, information is presently gaining access into the sense structures of the physical anatomy through his brain cells and presently reformed and stabilized in an integrated manner to give the individual so associated, a transcendency in what can be called a 'pickup'. He loses the weight of the associated nature of obsessional forms of his physical life.

On the other hand, an obsession can find entrance into the physical anatomy in the self-same manner as a transcendency. It is a dependency, hence an obsession, because there is a hindrance in the normal working relationship of the mental and superconscious circuits. This hindrance causes a malfunction of the energy wave forms carried piggyback, containing information that had been inputted from a previous lifetime, which had contained some negative quotients of a life experience, such as experiencing some psychological or physical trauma.

"That information reduces the energy of the individual, causes him distress, and instead of a 'pickup', he has a feeling of being degraded. As a result, he lacks confidence in himself, developing insecurity, distraught and guilt feelings. These inversions are indications of the malfunction of the energy, which is regenerating the defects exterior-wise, and now presently associated in such individual's life.

"And as we have indicated, since man lives on both sides of the equation of life, no man is singular and he is always completely embroiled with other human beings, and even nonatomic entities. He is being imbedded on the sine wave highway with the knowledge of other people's distresses and deformities. These are obsessional forms of energy that are the basic overlay within the fabric of any individual's life. It is an indication of a decline that has set into the mechanisms of such individuals who must function only with respect to their imagery of a physical anatomy and all of the various attendant sense structures associated with this physical anatomy. These are individuals who are living without the five tactile senses and who are forcing themselves on other people. In other words, they are intruding into the space of other individuals. They are broadcasting their selfish concerns into the mind of that individual and causing great static to his receiver, which is the sum and total of his consciousness.

"These individuals are unaware, in most respects, of this intrusion. Many are greatly aware and are purposely involving themselves in this fashion to dominate another person, and reduce his will. They can do this because of the greater strength which they have developed with respect to their previous lifetimes, lived in a leadership role.

"The sum and substance then of the obsessional forces and the obsessional nature of present-day life, is a proof again of evolution. It is a proof of the past civilizations that have declined, and have collapsed, because it was these very individuals who were the architects of these societies, whether Egypt, Greece, Rome, and all other societies presently known or unknown. The information that caused the decline is an overbalance and concern with the nature of the physical sense structure of life and less of a concern with the evoluting of the individual as a mind associated with a higher Infinite expression.

"Society is then the sum and total of the regenerating effects of energy, and man's mind carries within itself the features of its positive and negative associations with life experiences. For how else could it be? The positive features of life's picture in the individual and society is the movement of energy discharging itself from the infinite nature of man, his Superconscious or Higher Self, which has been created, developed and regenerated by each and every individual in his life to life experiences on the physical dimension of life, where he learns how to catalyze and synthesize his positive and negative experiences and distill from it the positive ingredients, which are the means by which he recharges the protonic nature of himself. He is then in this fashion, in this recharging, strengthening, building and regenerating the Superconscious circuit by adding more additional polarized information into it.

"But to all intents and purposes, what he has done is that he has polarized the present energy which is the substance of this circuit. He has polarized it with his own personal experience of life. It is in experiential living therefore, that the protonic nature of each individual grows when he learns how to discharge the negative materialistic nature of himself and to recharge the positive nature of the design of himself as an evolutionary participle of Infinite Intelligence.

"Alternately, of course, the mind functions in a charge, discharge manner in the third dimension to the fourth dimension by recharging the negative quotients of life; that is to say, giving greater strength to the visible evidence of the atomic form of life and to his ego picture of himself. To that extent, consciousness contains with it a greatly exaggerated picture of the subconscious. The subconscious is a substation only, and has given him the sense of being the sole architect of his life! In other respects, he has added to the lust for life in the power of taking life, by engaging in warfare for purposes of extending the boundaries of a nation, to advance the ego of certain individuals who are the power for that national entity!

"All is energy! The mind is an energy system functioning as an alternate current, charging and discharging the essential nature of itself in its evolution, as an evolutionary design of the Infinite. The mind can be obstructed in its evolutionary development through ignorance of its mechanism. Man was created by Infinite Intelligence, yet he can freeze his progressive evolution by holding on to the mirror of the atomic dimension and life in a physical body. When he departs the physical body at death, his hold on the illusion of his physical body is even greater as he lives with only one concern, to return to it so he can embrace the evidence of his five senses.

"It is these individuals, therefore, who are living in an adjoining dimension, unaware of the fullness of life that can be experienced in the fourth dimension, who are mutated by their negative frequency, biased from their previous life. This negative bias is therefore their means of vicariously experiencing the expression of a physical life. Obsessions are misshapen thought forms and their misshapen forms can clearly be witnessed on the features of individuals in times of great stress when anger and resentment are being transmitted into some other individual and visibly shown on the stamp of his physical form.

"This has been the experience of many individuals and will one day be shown visibly on a photographic plate or video image to indicate the validity of those who were experiencing a separation from their own internal intelligence and as a result have created a mutant form.

"Energy recreates the sum and substance of its own intelligence which is engraved indelibly on the electronic nature of its sine wave. This sine wave can carry the elements of the highest intelligence or of the lowest. It is man who is the selector and who is the adjudicator. He is also the judge and the jury, and he becomes his own headsman. So the mind contains the essential nature of life and the working of the mind is of greatest import."

Chapter 17

THE DNA CONNECTION

"Let us examine some propositions having to do with the enigma of DNA and RNA. These chemical compounds have, up to the present time, satisfactorily been resolved into the double helix, and as models, have been recreated to indicate the double strands of the molecule of the physical anatomy, which has revealed the basic architecture of the genetic cell structures of Homo sapiens. It is of interest that it is the large molecule that has been resolved so as to reflect its interior arrangement. And as a molecule is but the end product of many hundreds of thousands of atoms, the molecule is therefore not the complete picture of the primary energy structure.

"This primary energy structure is the electron wave form which, in its immensity of movement, has not yet been concisely determined by biologists with respect to the force that provides its structural design. This cellular structural design is the double helix known by microbiologists as the DNA, and the RNA chemical compounds which regenerate the basic organs of man's physical structure.

"These organs are, to all intents and purposes, large molecules that have been so interrelated and so associated that they provide an intelligent function for the generation of the mechanism that is the anatomy called the physical body. This anatomy, in every respect, is a basic mechanism that functions with the principles relative to Newton's Laws of Motion.

"However, the anomaly, with respect to all that is presently known in microbiology and chemistry, is the genetic code which is the regenerative feature of a cell, a protoplasmic cell that was initiated as a spermatozoon by the male. This cell is itself the most anarchistic factor in the concepts of molecular biology. And although the complete product is so revealed in each and every birth of any specimen, whether it be of a plant, animal or vegetable species, the transformation from a microcosmic molecule to a macrocosmic species of either a rose, a cauliflower, a giraffe or a man is, at the present time indeterminate. It is not complete in this transposition from the very tiny seed to what could be seemingly seen to be a giant mass of molecules such as in a giraffe or hippopotamus, or an elephant to that extent. But yet, the largest and most complicated association of molecules is with man himself!

"Hence it is that microbiologists and the entire field of chemistry are working with the third dimensional proponent factors of specimens of life, but because they have not yet developed this picture of the microcosm, that is presently the function of particle physicists, they have not yet been able to determine the origin of the atomic structures which are the basic determinants of the molecule. The molecule has therefore been carefully researched and has indicated that it has a very complex structure and intelligence in its complexity.

"Moreover, the staircase of the double helix indicates the direction that a particular cell or a particular molecule will develop into; what its fullness will be, whether it be a mouse, guinea pig, flower, horse or even a particular species of man himself, and in the many complicated sets in which the chromosomes of both the male and female have, in their pairing, regenerated. This is the basis of the present extent of research in microbiology.

"The attendant factors of this research are the chemistry of the various characteristics which are engraved within the biological specimen of that particular configuration. Therefore, the movement of all the parts, as they are functioning interdependently, are a function of the individual parts, such as the liver, the pancreas, the lungs, the heart, and the dependency of all of these organs upon the main organ, the brain.

"The complicated structure of the human anatomy is itself a visible proof of a greater intelligence than the sum and total of the parts of this organism. This intelligence, has not yet been determined, at least to the extent of its true nature. It has not been defined in a structural form! Up to the present time the only means by which any researcher in science has been able to formulate any understanding of the enigma of the third dimension has been through elemental means.

"Philosophers have attempted to explain the anomaly of man and of his physical anatomy which contains the basis of man's movement from some unknowable dimension. To reach some coherent understanding of the continuity from this unknown to the known picture of third dimensional life, mathematicians have attempted to explain the continuity of the mind through the abstract symbolism which transcends time and space.

"Psychologists have attempted to describe the movement of the mind of man as he lives in his physical anatomy, and explain the basic factors of behavior, both as an individual factor and in its sociological implication, in the movement of man within large groups, in different sets where the various types of behavior are formed.

"To all intents and purposes, religion is a recognition

that man lives beyond the atomic features of life, after the so-called death of the physical anatomy. But no satisfactory evidence has ever been given, at least within the framework of religionists, that such a life contains the intelligent factors that were the reflections of one's consciousness within the parameters of the third dimension.

"At the same time, neither of these proponents have been able to resolve the dichotomy between the mind and the body!

"Although on the one hand there is a great deal of information describing the interior behavior of the cellular organs of the physical anatomy, there is an inconclusive understanding of the architectural design, the blueprint that initiated the regeneration of its cell structures.

"To all intents and purposes, this is finally resolved with the realization that each atom contains a basic rendering of the interior of the structure of the anatomy in which the mind does function. With the addition of an interdimensional physics scientists will be able to describe the interior structure of the anatomy and give a detailed description of its electronic engineering, and the transmission physics of its design. However, scientists have not yet satisfactorily described the energy blueprint from which these interior designs are created!

"It is this energy blueprint which is the nature of the dialogues of this book *Interdimensional Physics*, to raise the perspective of man so that he will be capable of conceiving those mental constructs that have made possible his present appearance in the physical anatomy.

"In previous dialogues and dissertations of this and other texts that have been presented by Unarius, it has been indicated that there is an ever-developing feature, an ever-moving statement of the infinite of the fourth dimension, as it functions within all ancillary dimensions of itself. The ancillary dimension to which we refer is the third dimension.

"Since the third dimension is, to give a somewhat symbolic picture, a huge molecule, which is the greater of its parts. The molecule is the visible evidence of third dimensional factors of life. We mean that the mass artifacts of one's environment gives one the feeling of security and of having one's feet on the ground. This is due, in all respects, to the energy that makes possible the molecular features of atomic life.

"It is the underlying atomic picture therefore, which is the continuity of this molecule which provides its mass orientation in the third dimension. The atom is the foundation, carrying indelible factors of intelligence from its causal basis in fourth dimensional flux fields of energy. It is this fourth dimensional energy that is the blueprint for the combinations of atoms which, in turn, develop the complex series of molecules. And it is in this very complexity, that mathematicians are not yet capable of deriving any satisfactory solution because of the sheer numbers involved.

"A deck of 52 cards contains sufficient sets to make combinations of several millions of different sets. In the same respect, the hundred and nine known atomic elements in their association between each and the other, provide a greater complexity of sets which can be likened to molecules.

"Molecules themselves are the basic derivative from which the cell structure develops visible features, so that eventually, the organism as a whole takes on its mass determination, forming the five-sense structure of a physical anatomy.

"This great molecule that is the third dimension does not indicate the derivative of its intelligence through the evidence of instrumentation using the third dimensional physical tools, In other words, any research that is ongoing, is of the researcher himself. The researcher, in advancing the understanding of the sub-particle worlds that live within any cell, must also be aware that as he is an involved factor in the research and can only bring to the successful completion of such research, the awareness contained within the perspective of his higher consciousness.

"The particle physicists and molecular biologists viewing the subminiature evidence of particles being loosened from an electron or proton, would necessarily take that evidence one step further in the realization that the movement of this energy has been initiated from a region removed from the features of the third dimension. That is to say that the very atom that is being viewed is constantly divided, and will be further divided until it is discovered that the basic intelligence from which the atom had derived its power, was not from the divisible parts of itself but rather, from the cohesive nature of its own intelligence which does not live within the framework of the divisible molecular force of the third dimension.

"In this respect then, molecular biologists who are attempting to determine the basis of the intelligence for the structure of the physical anatomy and other types of species of life, cannot have revealed to them the bridge that makes possible this movement of energy from the fourth dimension which is of a higher frequency relationship of energy wave forms, unless he believes and accepts the inconclusive evidence of his present research to garner any satisfactory answers that this fourth dimension exists!

"Therefore, how will the existence of the fourth dimension ever be validated within the field of third dimensional research, except by those proponents who have evidence of this higher movement of energy within the nature of the avenues of their own evolutionary development. In other words, we see that any individual who was living in a physical anatomy and has lived in this body life to life and has not advanced into the recognition of the interdimensional nature of life, cannot advance the research in which he is involved, and expect to resolve this anomaly.

"The anomaly of the invisible basis of the fourth dimension has been explained, at least up to the present time, with physical instrumentation. Yet in order to understand the anomaly of third dimensional life, scientists must already have had a glimpse of this fourth dimensional flux field in their own life experience, previous to their coming back into the world of the five senses, from another world.

"And so the whole conundrum of life is that of the cat chasing its tail! Until scientists have themselves expanded their understanding of the infinite by having experienced it, they cannot possibly understand the higher nature of consciousness from another realm, because all people do live in another state of existence of consciousness, extended beyond the five-sense body!

"To the extent that they have developed the sixth and higher senses in their sojourn in this fourth dimensional world, they cannot have been or be capable of bringing this knowledge into their life or work when they again arrive through the womb into the third dimension on their return voyages!

"Of course this is not to move away from this dialogue! Without a concise understanding of each and every factor of life as a continuity from two dimensions, the individual cannot resolve the dichotomy and the enigmas of life! It is in this respect that even the most diehard scientist must come to the recognition, that there is no satisfactory relationship to an understanding of the mechanics of energy as long as he holds on to the *one-life-concept,* that he is more than the atomic mass of a physical anatomy!

"In this respect, the mass which is the evidence of the 109 atomic elements, is only a partial picture of the third dimension; yes, a partial picture! It is not a whole or complete picture, because these atomic forms are in their own resolution, functioning in their lower frequency as energy, yet a factor of the higher polarity or integer of which they are a part of their isotopic nature. In other words, the mass of the 109 atomic elements symbolically expressed by M in Einstein's theorem $E=MC^2$ is a real factor of the energy "E". Mass as energy can only be fully understood when it is multiplied by the constant speed of light, squared. The true configuration of the atom will then be understood as pure energy which does not function within the proposition of the constant speed of light of 186,272 miles per second, but is an additional product of infinite intelligence or of Infinity.

"It was this very understanding initiated by Einstein that can be taken a step further in resolving the mystery of life and so-called death. How is it that at one moment, we see a virile physical body moving and accomplishing many complex aspects of life within the proposition of some individual's basic intelligence; then at another moment, we see that very anatomy stretched out on a morgue as a corpse, having no sign of vitality, inert, containing a minus of intelligence that had enervated and moved the many organs of the body, but particularly, the brain?

"It is this very factor of the atomic mass of the physical anatomy that can be equated to the large molecule which we have used as a symbology of the third dimension. For mass cannot function without the attendant factors which gave it its basis for life; that is to say, of intelligence, because the body is not intelligent when it is lifeless. Therefore, that body is basically a mechanism that has lost its power!

"Where is the power? What is that power? And what are the characteristics that make possible the functioning of this power so that it is rectified within the physical anatomy? This is the entire proposition that will finally be resolved in the understanding of the mechanics and property of the mind.

"It can thus be seen that the corpse contains within its cranial structure the gray matter called the brain and that the *brain is not the seat of the power that activates the physical anatomy*, else that physical anatomy would not be a corpse!

"Hence, taking this a step further, as have other intelligent researchers into the state of life's existence, there must be a power that functions from another dimension, from another source! It is therefore this source, that has been proven to exist; because its existence is mandatory in the continuity of the physical anatomy whereby the physiology of this anatomy is recreated or vitalized, and is a viable force.

"The DNA therefore explains only the regeneration of the cell structures which make up the basic design of the anatomy. The mind therefore must be the basis for the deeper research into the causal factors which are attendant upon the movement of these organisms. This mind is, in this respect then, the basis of the research into the attendant fourth dimension, because energy is a factor in the movement of all visible and invisible aspects of life.

"Energy, as we refer to it, that functions without the dependence upon the mass of third dimensional life, is the factor of a fourth dimension because its origins are not within the molecular nature of third dimensional cellular structures. Hence, the mind is the only means by which this energy structure can be determined. The mind is then the intelligent design that contains the evidence of the existence of an energy field that functions above and beyond the third dimension, and yet functions in some manner whereby it can be perceived and detected in the third dimension. It is this anomaly which is a visible factor in the life of all human beings and gives him the opportunity, at least to this extent, to come to this conclusion through the analysis of his own life experience.

"The *mind* is therefore the *bridge from a higher structure,* a radiant field of energy that is unseen by and through the factors of the five senses. It will be the need for man, therefore, to develop a higher sense structure in order for him to detect the presence of this fourth dimensional structure called his mind. Hence it is, that research in the developmental features of man's life is being directed into the learning and the application of these higher sense structures that are the nature of his energy field of the sixth, seventh, and higher senses. This provides the evolutionary proof of life in the recognition of the paranormal factors of life beyond the so-called norm, meaning the five sense features which are indicative of the continuity principle of life.

"Moreover, the readily recognizable logic applies that since the mind has been a development of fourth dimensional fields and that the unified field contains intelligence, it provides an ability for each individual to apply such information as he is aware of in his life-to-life evolution, and which are now products contained in the mind.

"Logic and reason are capable of transcending and thereby overcoming many factors that are engendered into life experiences and interrelationship between one-self and his environment made up of people, places and things! In this respect then, man is carrying within himself the dichotomy of life. The enlightened researcher into these dichotomies, who has armed himself to a greater degree with certain proponent knowledge of the movement of energy in its lower form, cannot help but realize, with the application of logic and reason, that his mind is nonphysical to the extent that he is able to transcend the features of which he has made his mind a factor of intellect!

"Let us discuss man's acceptance of the mass properties of life as being the sole factor of life. When any scientist is able to transcend the mass factor, then he is on sure ground and will eventually be able, in his special discipline, to resolve the paradox of the nature of his particular expertise in his research efforts into the present disciplines of life, whether it be in physics or in philosophy; whether it be in music, psychology, or in other factors of life having to do with the understanding of planetary life and its origins!

"No matter what area of life research is presently being directed into the understanding of the beginnings of life and the continuity of it, it will finally come to any individual that he holds within his own mentality certain basic keys. Those keys are the evidence of his own continuity from life to life, and the evidence that he has lived previously and has been carrying the basic propositions of his own present knowledge from the many hundreds of thousands of lives that he has lived!

"If he should be in a condition where he has not accepted the basic sine wave principle of life existent within and functioning in each individual person as his continuity in the great evolutionary movement of the infinite, then he is stalemated! He is not capable of resolving such enigmas - one which was indicated by the French mathematician Fermat whose famous 'Last Theorem' has to do with the nature of solving the dichotomy of man's ability to function beyond the third dimension and still contain the consciousness that had been so rectified within the proposition of space and time.

"The causality of life as it is lived by any individual cannot be resolved in the mathematics of third dimensional physics. This is because that individual who is held within the restricted boundary lines of his own theoretical determination and illogical symbolism, can never transcend that wall and see behind it.

"If it is true that Fermat had proved his theorem of 'X' to the Nth power, 'Y' to the Nth power, equals 'Z' to the Nth power, he was not a third dimensional scientist! He therefore understood the proposition that was posed, using the mathematical symbolism of calculus, but he was already capable of transcending the enigma posed by the theorem because he knew the answer. Because the answer was already an established factor of his mind which had lived in the fourth dimension, he therefore had recognized the continuity of energy as the continuity of the individual and of his infinite perspective of life!

"Therefore, the fact is that the third dimension as a spectrum indicates a separation from a fourth dimensional continuity. It is only the inability of such an individual to see beyond the mirror reflection of life that is constructed in time and space, a beginning and end, an up and down, life and death, and all of the other factors which are indicated by the proposition of the sine wave restrictions within certain lateral boundary lines.

"Only an individual who has already developed these fourth dimensional equivalents, could possibly advance his understanding within the underside of the Infinite, to recognize what lies on the other side of the curtain.

"The only answer to explain such enigmas is to traverse the square and to move out of the mathematical symbolism that has been dependent upon third dimensional logic. Thus, in one fell swoop, there will be the realization that all of the theorems proposed inquiring into the integrated understanding of energy have been propounded because these individuals had already received certain awareness while living in the unified state of consciousness in a fourth dimensional world! Depending upon the successful development in the discipline in which such individuals were students of life, they could advance an understanding of certain abstractions which they had deciphered and apply this intelligence to the lower axioms occasioned by their association to a physical anatomy.

"The atomic nature of life then, is the basis of the move forward into this *Age of Logic and Reason.* Logic and reason contain the answers to the puzzle of the circular nature of life. It contains the answer not only in the syllogism drawn on the blackboard, but it is contained within the individual's own awareness of himself as being this proposition. It is in this factor that life so advances, because the advancement is being made possible by those teachers who pushed the boundary line of life further and further out from its obsolete design.

"Hence, those individuals who were responsible for advancing the awareness of the subliminal nature of the intelligence residing within the physical anatomy as the DNA code of the double helix, have made possible the continuous detective work which is the nature of microbiology into the avenues of the supporting structures into the atomic nature of the molecule.

"In the work of biology and microbiology, the electronic nature of the cell structures will be advanced and chemical reactions will then be recognized as an electro-mechanical movement functioning to maintain the blueprint of the DNA structure. This energy field will be finally resolved as micro-energies constituting the surrounding flux fields which is the prime architect and is constantly projecting its intelligence for the continuity of the life of the third dimensional construct of the physical body.

"This energy field which surrounds each cellular structure in any characteristic organic or inorganic material will therefore be recognized to exist in its regenerative fashion as a flux field which has already been indicated as a great electromagnetic field, such as can be seen surrounding any planetary body or sun. This electromagnetic field which has been given the name, the *Kirlian Effect,* is the evidence not only of Russian researchers but also by American researchers.

"It is this great flux field that, once removed from the cellular reflection of itself, causes the completion of the physical life of that particular mass organ - man, animal, vegetable or mineral! Hence, the third dimensional atomic anatomy can only contain its life if the electromagnetic field, which is its first dependence, is continued, for it cannot live without it!

"We see in this respect, the similar coupling of the television or radio transmitter with the television or radio receiver. If there is no feature of activity on the television receiver or in the audio reception of a radio receiver, then there is no life being inseminated or projected to these receivers! The physical anatomy can be considered similar in this respect to be only a receiver of intelligent energy from a transmitter. The electromagnetic field then, that surrounds every organic and inorganic structure in this third dimensional spectrum, must therefore contain itself only from, and dependent upon the flux fields of the transmitter. Therefore, the macrocosm that makes physical life possible is dependent upon its transmitter which contains the causal nature of the information activating the atomic picture of life on a third dimension.

"That transmitter is a field force that is receiving information of a fourth dimensional nature because it resides upon a higher frequency level. This then is the proof of the continuity of all third dimensional forms of life as it has its basic dependence upon the transmitter, transmitting from a higher flux field of energy, and becomes the diagrammatic nature of third dimensional life. Therefore it can be seen that the third dimension is a reflection of the fourth dimension! It can be seen then, that the physical anatomy is only a reflection of the architectural features that are the blueprint of the mind.

"The mind can hence be seen to be a factor of fourth dimensional development. It is the mind, when its mechanics are developed, that will give a greater and greater rendition of the validity of the Interdimensional Cosmos and resolve, once and for all, the superstitious religiosity pictures presented by half knowledge and half truth. It is only through science, but more properly, fourth dimensional science - pure science - that the physics of the infinite will be brought into the framework of the consciousness of man while he is yet living in his physical anatomy. This is so, because the physical anatomy is a developmental factor in the beat frequencies that are the coupling of energy wave forms oscillating from polarity to polarity.

"The entire nature of life is therefore a constant climb to gain a wider perspective from the additional information inputted into the transmitter. The transmitter thereby projects a higher resolution of a wider band sweep. Therefore, the interrelationship between a transmitter and a receiver is a constant recreating of information to develop its own electronic nature.

"Let us therefore see that the entire nature of the subdivision of knowledge that is the present disciplines in the educational format of life, will reveal from one to the other, the basic truths of the interrelationship of all factors of life, from the works of the molecular-biologists, the microbiologists, the particle physicists, the astronomers and cosmologists, as well as from the works of the sociologists, the physiologists, psychologists, geologists, anthropologists, and so on and so forth.

"A clear picture of great order and intelligence will be the resolution of these fields of inquiry as they are coupled together in the *New Age of Renaissance.* Their coupling will provide the knowledge that it is man's mind which will be the next step in the resolution of science, the resolution of life and, of course, the resolution of mankind as he is becoming aware that he is a giant molecule who is just beginning to recognize the nature of his atomic constituency.

"Fourth dimensional physics is not a proposition that is dependent upon the symbolism of parts of knowledge that have been so insistently taken as to be the complete answer to life. Many individuals have misappropriated and misread the basic principles that have been advanced in previous lifetimes by some advanced Intellect who contained the fourth dimensional proposition of life within himself.

"Man must be dependent only upon his own logic and reason, which can be gained by the knowledge of the *physics of his mind.* It is in this great field which is the reflection of Infinite Intelligence whereby the truth can be attained and applied by any individual. It will be in this fashion that he will be able to trim his sails as he crosses the vast open seas when he leaves his physical body and journeys into the fourth dimension to add to the knowledge of his own purpose and objective in life as a *Light Bearer,* to carry to the four winds, so to speak, the knowledge of a Creative Intelligence."

Chapter 18

MAN, THE ELECTRONIC COMPUTER

"Present information attributed to research in neuropsychiatry, of the nature and function of the brain has indicated that the hypothalamus and pineal glands contain a link to the higher functions of man. Of the four main lobes of the brain, the cortex is considered to be the seat of the rational thought processes! An immense amount of data, baffling the most advanced computers is correlated, and selectively determines the basic factors that make possible the processes of living by any individual.

"These many complex bytes of information are constantly reflected from the circuitry of a non-atomic source, and contain the memory of each individual's past life associations. These life experiences are reinstated in the present, and determine the emotional and intellectual responses relayed through the central nervous system of the physical anatomy as speech, sounds emanating from the larynx to form various signals into patterns called words.

"Yes, the autonomous and cognitive nature of the brain is the mirror of the mind, as the brain is the negative polarity of the electronic body or the psychic anatomy, an anatomy that like a computer engineers the entire design of each individual's evolutionary development. It is this oscillatory movement of intelligence, that is the mode of Infinity.

"The Infinite is reinstating itself into a lower frequency movement of the interdimensional cosmos and regenerating through third dimensional modes of expression into cellular structures. A fetus is the regeneration of energy information from non-atomic structures, demodulated to a basic frequency of atomic constituency, to recreate and regenerate the mass organ known as the physical anatomy.

"It is the brain that is the negative polarity of man, a transforming device to regenerate the intelligence of the positive polarity, the mental body. As we have indicated, the mental body or the electronic anatomy is constructed of three circuits. These three circuits are necessary in carrying out the evolutionary nature of life, to provide each individual the prerogative to learn how to integrate the various life experiences which will determine the basis of his evolutionary development. The regulation of information carried out through three basic circuits reaffirms and reestablishes the continuity of man's consciousness with his present physical state in his awareness of life.

"Now that we have set more into bas-relief, the awareness that man is a relationship of two associated fields of energy, we can substantiate this statement Einstein's mathematical formula $E = MC^2$, where 'E' stands for the energy body of man, his electronic anatomy; 'M' stands for the physical anatomy, and 'C' stands for the speed of light squared. The electronic anatomy is intricately associated with another energy field, oscillating on a higher frequency, dimensionally-wise. The mass body of man is encompassed within the light factor of a third dimensional field in which all mass bodies are so associated.

"However, man's true association is with the non-atomic body, which is not third dimensional, and therefore is the more important in understanding the physical anatomy. In other words, the physical body is a derivative of the non-atomic energy body. With this geometrical analysis, there is the possibility for a greater logic for any individual scientist, that will explain the enigma confronting him in his scientific research.

"The three basic circuits of the electronic body of man provide in each lifetime the potential for man to gain newer perspectives of the nature of his own energy system. Such higher perspectives are therefore based upon the attribute of man's higher energy nature. Consciousness, which is the basic note in the understanding of any individual, is a closed circuit, a field force oscillating from mental to the physical brain, a beat frequency between two opposite polarities!

"The mental brain contains the complete diagram constructed from electronic energy sine waves and functions as the main transmitter. The physical brain is functioning as the receiver. However, we hasten to add that since this polarity relationship between the two opposite poles of man is a regeneration from positive to negative, there is a reversal when the negative pole exchanges its information to the positive pole. Therefore, the transmitter and the receiver are also transceivers!

"Man is a receiving and sending instrument! The physical brain of man is a receiver and a sender. In this respect, we have the basic outlines of the electronic system of man's consciousness. That is, the functioning nature of man's present consciousness is related to him from a non-atomic field to an atomic field through man's physical brain lobes. The atomic field energy system, in turn modulates and effects the non-atomic energy system of the mind.

"Therefore, Consciousness, which is the expression of life, is a closed system that is receiving its information through the transmittal of program data that has already been inputted and is a factor of the energy body. For the sake of simplicity, let us call this energy body the psychic anatomy. It is the repository of all information that has been inputted into it. And since the input has been generated from the movement of energy, life after life and interdimensionally, the computer contains all signals that have been the accumulation of the life experiences of one individual from the imaginable past, stretching back millions of years earth time.

"This computer, you must understand, is fourth dimensional in its design, a circuit containing the entire nature of an interdimensional intelligence. The higher carrier wave form structures are demodulated to an atomic frequency so that each individual can experience life through a physical system that is finite.

"This is similar to the demodulation of the solar energy of the sun into the primary and secondary wave radiations received on earth. Therefore, man in his initial spiritual evolution, learns of his *Infinite Intelligence* through the primary frequencies of a lower energy. The higher cellular energy circuit is the carrier wave regenerating itself into its countless finite forms, to become the beginning of man - if we can call it a beginning. Yet, it is a beginning in which man initiates the constructive evaluation of himself by acknowledging the prime nature of his *Infinite Architecture.*

"This energy field is non-atomic and contains within it, the entire DNA structure that is associated with the physical anatomy; it is the architect of the biological system and other factors involved with the mass of the physical anatomy.

"The mental body is in every respect the origional or radiant energy body, providing the transmissional basis for the physical anatomy in terms of its cardiovascular, pulmonary and brain system, etc. In other words, the entire shape and form of the physical anatomy conforms to the picture of this mental or energy body. Hence, the physical anatomy, in all respects, attains its form and structure from the structure and function of the mental body.

"Many individuals have alluded to this as being the astral or light body. It can be considered to be a body made up of light, and in fact is a body that is formed from the radiation of the higher isotopic atoms of the physical anatomy which are non-atomic in their oscillation.

"It is the physical anatomy which is rendered from the astral body. Therefore, in order to learn of one's true spiritual body, one must develop the understanding of self through the development of a lower organism which grows according to one's development and ability to function in greater and greater complex fields of energy wave forms.

"The mental body is a function of the fourth dimension, and because of the frequency of its energy structures, it has a libido and drive of a higher potential, in-phase with an Intelligence which is the nature of this higher potential. In other words, it is unified in its awareness of its spiritual nature.

"Intelligence is, in every respect, the perception by an individual through his selective abilities to objectify his relationships with the alternate poles of himself. The alternate poles of man are therefore, in a larger reference, the movement of his fourth dimensional body, his positive polarity, to a third dimensional body. which is his negative polarity.

"Since Infinite Intelligence is correspondingly moving in its oscillatory fashion, positive to negative, its negative polarity, shall we say, in this great cosmological statement, is the third dimension. Therefore, the spark of this Infinite, which is man, must correspondingly oscillate with its negative polarity, which is the third dimension.

"Evolutionary-wise, man is learning of his interdimensional nature through his involvement in his life experiences on the lower structure of the interdimensional vortex, the nucleus which contains the equivalent polarities of the interdimensional vortex.

"The nucleus is the compact atomic energy flux field that has been pushed down by the greater pressure of the expanding centrifugal pattern of the interdimensional vortex. Mass, then, can be seen as a dimension of atomic energy, as a demodulation containing the negative polarity of the positive factor of itself, which is the fourth dimension. Man, hence is presently functioning as a polarity relationship of the Infinite in its nuclear form.

"The need for a physical anatomy then becomes quite paramount; in fact it is mandatory in the understanding of the interdimensional physics of the science of life. Therefore, the physical body contains the opposite structures of the fourth dimensional factors of life, refracted into man while he occupies an atomic body.

"Now all enigmas associated with the metaphysical teachings of Buddha, Osiris, Akhenaton, Krishna, Vishnu, Socrates and Jesus, etc., become quite understandable. Such enigmas reiterate the need for man to realize that the atomic physical body, the mass third dimensional body is only a means to an end - a learning device. And although that physical body contains the reflective picture of the Infinite, it does so in a closed circuit, which does not give the entire nature of the evidence of a greater grid of the cosmos, because its picture could not be presented on the small screen that man is capable of viewing. He does not have the ability, yet, of the knowledge of his electronic nature and the infinite perspective of life.

"In other words, infinity is incremental as is man's ability to conceive of it. This may seem a contradiction to some people. It was never a contradiction to the great Teachers and Avatars such as Jesus of Nazareth, Socrates, Buddha, and the Chinese Sages, Tai Ching, Kung Fu, Mengtse and others. They had reference points for explaining of the separatistic nature of life viewed through the physical anatomy. They knew about the positive polarity of man and spoke of it as his mental self and-or his spiritual self, for it is truly one and the same in different allied relationships.

"Therefore, in order to function with a physical anatomy, such physical anatomy must contain the properties of another primary anatomy, because the physical body is only a reflection, an atomic mechanism that functions only from the power transmitted from an initial electronic device. Therefore, the transmitter itself must be a containment of energy of higher circuitry, which is non-atomic and fourth dimensional body.

"The subconscious contains the negative polarity structure of the electronic anatomy. It is labeled subconscious, because it carries within its data bank the associated information of man's past-life experiences. That is to say, it contains all information relative to the emotional contents of life lived in and through a physical anatomy on a terrestrial plane. The subconscious makes possible the regeneration in every respect for the secondary body, which is properly associated with the fetus of a beginning physical anatomy.

"The entire proposition of birth and death is contained within the nature of the constant oscillation between the circuits of the psychic anatomy and the physical anatomy. Life cannot be deciphered except with the understanding of the relationship of the corresponding movement of electronic wave forms from polarity to polarity, pole to pole. It is always in this respect that we have intelligence. Intelligence is the sum and total of this oscillation between north and south pole. What the result will be is the frequency associated with that oscillation and the interactive movement of the wave forms as they have been recreated by harmonic factors of each individual associated through his countless life-to-life experiences.

"The physical body cannot be constructed unless there is a polarity involved. In this respect then, the individual who is living in his astral or light body, his non-atomic body in an adjoining fourth dimension, must have a polarity that is situated on the negative side of this sine wave movement of life energy. This individual, therefore, who is properly placed on the negative polarity, makes possible the recreation of additional life experiences. In fact, it is necessary him to gain additional experiences from the negative polarity of himself through his mass sensing device, the physical anatomy. Therefore, the entire process of birth begins with the coupling of two individuals who, through the regeneration of the polarities of their cellular structures, through the spermatozoa, regenerates a physical anatomy.

"The cellular mass body has the ability to grow or to regenerate through the process of mitosis; yet it does not have a basic life force that has a cognitive ability. Man's cognitive nature is the property of a higher frequency energy structure, which is the mental body of any individual who is preparing to oscillate to this physical cellular structure, negatively biased, to a terrestrial plane. Therefore, through this means he can oscillate and establish a relationship based on the understanding that all energy must move from one polarity to another. Thus, as he is posed in his positive polarity, to all intents and purposes, associated with his non-atomic body, he can begin to express the nature of his developing consciousness to the extent of what has been learned from his progress in life from the negative plane of life on a terrestrial dimension. This development has been attained through his ability to stabilize his electronic body, which oscillates in and through the cellular structures of the physical anatomy.

"This physical anatomy is a regeneration of base plane frequencies of atomic configurations. The fourth dimensional thinking body, shall we say, the body of the transmitter, provides the physical body with the necessary motivation for the association of this individual as a continuity from previous lifetimes, in which he has occupied a physical body. In fact, every individual has the familiarity of his physical body. because of his previous experiences with it.

"The mental body is therefore transmitting into the physical anatomy all of its necessary cellular attributes! The mental body, in other words, is the *architect* which forms the form, color, size or whatever, which becomes the recognized picture of the physical body. This biological structure is a replica of the astral body.

"The subconscious body is the demodulated energy structures of electronic energy wave forms from the mental body so constructed that it provides a complete contour map in which this physical body can be regenerated.

"Each life that man or woman lives is a life which was constructed from the subconscious circuit of the psychic anatomy. That circuit contains the entire picture of that particular anatomy, and is therefore a different anatomy containing the biological factors of its composition from one life to the next. The harmonic affiliations which are the continuity of the memory of the individual from his past, are integrated into this body, which contain the sum and total of all experiences that have been lived by any individual in different physical lifetimes.

"This is the means by which a new physical anatomy can be grown from the harmonic affiliations of the genetic structures of the parents, father and mother. As the child sets up his oscillation with the negative plane of a physical body, he adds an additional factor in the biasing of the characteristics that make up the generic nature of the body as male or female. The individual therefore contains all of the attributes which are its DNA structure previous to the construction of the physical anatomy, which biases the sexual nature of this physical anatomy.

"The physical brain structure is the demodulation of the mental body of man. The brain contains in its complex oscillation of energy wave forms, a replica of the brain of the non-atomic or energy body. The brain of man is a cellular receptacle and replicates the nature of the man's intelligence, which is the accumulated evidence of such individual's evolutionary development. In other words, it contains the nature of all factors of life in which he has processed information and has determined that this or that either works or does not work, to enable him to advance his life from a physical to a mental development. It enables him to build houses or to build dams, to irrigate the land and construct power systems to provide for the improvement of living in cities or in farms. This is the constant improvement contained within the ability of any individual to rectify the electronic processes which are the functioning attributes of his Superconscious Self.

"His superconscious self, as it is a part of the Infinite, is expanding, because as we have indicated previously, the infinite is the aggregate of finite factors. Each individual finite factor in its association to other finite factors, make up the infinite. Therefore, as one atom is associated with another atom, its molecular structure will contain the matrix that is infinite in correspondence to its individual unit. Three atoms, four atoms, and a combination of hundreds of millions or billions of atoms, will in this respect, make a larger picture of life. Incrementally-wise, these are finite factors which have been enlarged through association, integration and intelligence.

"Therefore, the third dimension can be considered to be a finite attribute of the Infinite. But at the same time, it is one of the alternate polarities of the Infinite. The positive to negative polarity oscillation is a constant and incessant movement within the flux fields of the electromagnetic seas of energy!

Life is constant motion. Birth is a motion from the fourth dimension to the third dimension. But again, death is a motion from the third dimension to the fourth dimension. Death is now birth in the fourth dimension. The fourth dimension surrounds everyone. Ernest L. Norman has indicated how the infinite is at the tip of everyone's nose. It is invisible for the vast majority of people and therefore, does not exist. Since the physical sense factors of life can be considered to be the reality of man, the dissipation of these same sense structures has instituted the belief that the physical is the completion of life, or the death of life for man.

"This belief is itself a dichotomy, because life can never be dissipated! The recognition of such incongruity to the sensing organ of a Homo sapiens, who is a soul of the Infinite, is an indication of the breakdown in the oscillating movement of man himself, occasioned with ignorance or the lack of consciousness of the ever moving omnipotent, omnipresent Infinite Intelligence; in that such an individual is not functioning or is not in tune with the positive polarity of himself to the extent that he becomes conscious of his own developing Infinite Intelligence!

"In other words, man is unconscious if he is not in tune with and functioning with both sides of his polarity nature! He is asleep although he is aware through his five sense structures, of the physical mass reflection of light! But he may be asleep to the non-atomic dimensions, yet he is awake to a non-atomic dimension in a so-called sleep state, in tune with the positive bias of his psychic anatomy. Of course, we must remember that the psychic anatomy contains all references to the associated life experiences that had been distilled from past-life experiences on the physical plane of life, as well as on the spiritual or non-atomic plane.

"But for the most part, the majority of man lives biased to the negative side of the Infinite or the third dimension. Therefore, he is third dimensionally attracted or biased with the understanding of life from his previous earth lives. And to the extent that any individual has grown and become aware of his positive polarity, the higher mental and spiritual Superconscious nature of himself, he has developed the higher sensing signals or circuits of himself.

"In this respect, when these individuals are in socalled sleep state, they are attracted to the fourth dimension from the perspective point of the higher sense reflections, because the fourth dimension has many increments within it, as do all dimensions.

"As man moves from his negative polarity to his positive polarity or from his third to his fourth dimension, from life in the physical body to life in the non-atomic body, a development is constructed and shaped by the design of his psychic anatomy. When living on the other side, as it has been called, on the astral polarity of the physical life, man is living with the memories of past associations with his physical body, which were constructed from the images from his physical statements of life; that is, from the reflection of the physical five-sense body. Not having learned what is surrounding him, now that he is without a physical anatomy, one is still attuned and in-phase with past memories of life lived on the Earth. These are the lower frequencies, biased from the negative polarity of consciousness.

"So this means that the development of Intelligence, the development of the physics of man himself, is a continuity over many associated lifetimes, because one life is a very brief span. And the return tour that an individual takes can necessarily cause him mental retardation if and when he does not learn from his experience. He then repeats and is forced, in this respect, to repeat the grade in the school of life, causing greater degrees of difficulty, because he has added more of the negative reflection of the Infinite which has not been balanced with the positive nature of himself. In all respects, it is in the movement from the physical life to the spiritual life that provides an individual with the strength of each opposite polarity.

"To all intents and purposes, the positive pole contains the higher incidence of the engraved Intelligence of the Infinite, because it is a standing wave of a higher frequency! Standing waves are the parallax of the electromagnetic flux fields surrounding a physical planet, as well, of course, in a spiritual or fourth dimensional aspect, in their parallax, associating with an integrated interrelationship of information within the structure of each of these electronic wave forms.

It is very much like a satellite orbiting an earth planet, whereby a laser beam can be reflected, because it is orbiting at a certain distance in its ellipse about the planet.

"A standing wave acts as a reflector shield. Man, in this respect, is very similar in movement on his life-to-life base plane frequency. Experiences which have been parallaxed through their lines of force, contain information which has been the developed nature of information now in harmonic attunement; they have a harmonic affiliation. This information may be either negatively or positively biased. It is this association in the movement between the two polarities of what has been called birth and death, from adjoining dimensions that a standing wave is developed; the sum and substance of previous associated life experiences.

"Therefore, that life experience, such as an acute trauma from having died on the high seas in some excursion from one continent to another, is energy in sine wave frequencies, oscillating within the electronic body. Such similar experiences are associated to form a specific frequency as a standing wave within an energy structure, which has been recreated through life experiences on a terrestrial plane.

"The standing wave is the trauma of dying in a great hurricane, where the individual was cast from a ship into the boiling seas. This will be an immutable memory which he will be reattuned to when he in crafts a new physical anatomy to continue his evolutionary development in the learning of self.

"The psychotherapy factor of life is the rectifying of various traumas of life in the innumerable interrelationships of any individual with other individuals who are functioning in different frequency and harmonic relationships.

"Everyone, therefore, has formed certain standing

waves which are reflected directly back to the individual in a similar manner in which that individual has lived in some previous life, thousands of years previously.

"The energy of a standing wave is not atomic. A standing wave is a field of energy that contains the entire history of an individual, which is the product of the accumulated information which is presently the video tapes being instrumented by his computer body the psychic anatomy - into the present body to which he is alive.

"Such an organ can be a physical body or an astral body, or other bodies that he will rectify and live within as he progressively moves up into the higher scale of the Infinite Cosmos. We have been taking you into the higher constructs, explaining the physics of Interdimensional Science.

"As the majority of people at the present time are searching for their spiritual polarity, through the evidence of the physical sense structures, they are, in every respect, still functioning within the negative polarity of themselves. They have been able to reconstruct and gain certain evidential attributes of this physical sense, mass structure; that it contains factors within itself and cannot be determined by separating its atomic element.

"If a physical body were to be placed upon a table so that all of the organs were removed, and each cell of these organs was separated one from the other, there still would be no ability to determine the origination of the power that fibrillated these cell structures. The power you see, as you have no doubt guessed by now, is not sensed entirely by the instruments of the analysis. The power for the movement of this flesh and bone body is the positive polarity. It is a power that cannot yet be indicated by physical instrumentation capable of sensing the evidence of the molecular nature of the activity of the cellular structure.

"The electron microscope had not yet been able to reflect the atom and particle accelerators have not been able to capture an electron, so how in the world is it possible for man to realize the incredible movement of the energy that is the atom? He cannot possibly freeze this movement! But why can he not freeze this movement of the atom? Because it is moving faster than the speed of light! It is moving beyond his capability to rectify this frequency into physical consciousness.

"Conceptual understandings are attained from the rectification within the cortical nature of man's brain structure, where he determines the fourth dimensional nature of his mental self. That is to say, the mental self is an instrument for man to see the fourth dimensional movement of himself and to validate its appearance in his life.

"But until the understanding of the polarity reversal of energy is developed with respect to man, he cannot possibly understand the world objectively. The object is so intricately related to the subject that there cannot be the objective attitude of logic and reason in life until man becomes aware that he is a reflection of himself. He is his own apparent reflection as he puts his eye down to the eyepiece of the microscope or to the telescope, for he is, in every way bound up in what he sees, because he is also seeing a factor of himself!

"So it comes down to the basis of man to 'know thyself'. The proper basis of research is man himself and that research must take man into the nature of his electromagnetic nature! The electromagnetic nature is this complex movement of his mind, and the understanding of the mind will resolve all enigmas.

"The Unarius Science is a futuristic science. It is a natural science. It is a science of life, in which all individuals are a functioning part. It is the diagram of Infinite Intelligence!

"In our next chapter we will discuss the basic mechanics of the brain. We will take you on a journey into the past and to the future. This is the nature of the plasma which is the exchange between each individual cell of the brain and its transmitter. We will show the delineation between the fourth dimensional brain and the third dimensional brain, to highlight the highway of life that is each individual's continuity from his past to his present, and from his present to his future. We will indicate that the sine wave principle of life is functioning in every nanosecond of consciousness!

Chapter 19

TOWARDS A PHYSICS OF CONSCIOUSNESS

"The total concept of the Infinite Cosmogony as a personal proposition to learn and to understand at least to a certain degree, is indeed, without question, a mind-boggling proposition. All peoples, as they have so evolved on the planet earth, have conducted or incurred the total consensus of this evolution on a strictly third dimensional basis, and except for certain inspirational values or occasional paranormal phenomena, the earthman has only recognized this interdimensional cosmos through such vagaries as superstitious beliefs, religions, etc., and has to some degree, succeeded in strengthening such beliefs through the necessity to relieve and alleviate the fear of death, always incumbent in this mortal life about us.

"The psychodynamics of mind as an electromagnetic structure establishes the nature and reality of consciousness as an interdimensional energy process. Diagramed in the language of physics and electronics, this New Science clearly explains the enigma of physical and mental disease, obsession and possession, hypnosis, telepathy, out-of-body experiences, past life regression, poltergeist, and the extraterrestrial phenomena, etc.

"Consciousness can be described as an electronic process of rectification and purveyance, expressing in the polarity function of cause and effect. Past-life experiences are 'programmed' as information energy sine waves and are contained in the master computer or psychic anatomy to be reinstated life-after-life as consciousness. The incarnation of an individual at birth is the beginning of a new physical cycle and death is the completion of this cycle. In this respect, one's perception of life, is one's intelligence or concepts regenerated life-after-life, positively or negatively biased, forming the foundation for present consciousness. A greater knowledge of the electronic physics involved is mandatory for the correction and cancellation of any malfunctions expressed in the breakdown of the individual's relationship to society.

"It is man's prerogative to extend his consciousness beyond the limitations of the "five-sense physical anatomy", to explore new potentials within the logic and reason of progressive evolution by the development of the paranormal facilities of the mind. The resultant attainment of an inter-dimensional or cosmic consciousness will have an important effect upon the global society of man!

"All the familiar earth forms including our own physical anatomies are reducible to those energy forms classified in scientific nomenclature as atoms. These atomic forms universally, so far as the third dimension or the terrestrial world is concerned, form a surface plane or a curtain, if you will, of energy upon which surface man is beginning to learn the combination of factors not heretofore relevant in the more primitive stages of his evolution. In other words he is learning to combine and recombine these atomic forms as the beginning of a facade of life which will later conceivably and predictably carry him into higher or more fully-developed planes of life, where all such relevant surface plane expressions, will be quite different than present third-dimensional forms. This is the basis for a progressive evolution, which is the learning of the psychodynamic properties of the Mind!

"This mental process reveals the ultimate nature of each individual's thoughts as information sine wave energy oscillating their positive and negative polarities from specific life experiences, compounded through previous lifetimes. Electronic energy wave forms, as complex harmonic configurations, ring-in the accumulated nature of all life experiences according to the specific frequency from which these experiences were recorded on the master computer or psychic anatomy.

"Therefore, every thought expressed by any one individual contains within it the so-called mystery of life! It contains the origin of that thought in time and space, when it was first expressed. It also contains information of life experience from a timeless, spaceless dimension, lived on non-atomic planets. The brain cells as terminals serve to demodulate the information energy wave forms as thought which is then amplified through the five sense organism of the physical anatomy.

"This is known as consciousness. Therefore, we see the importance of this understanding of the highway of the mind as it provides a contact between the terminal, the brain, and its master computer, the electronic body. The master computer is an electronic configuration that is contained in three circuits, through which it establishes an equilibrium for the individual throughout his lifetime. Because this movement must always be an oscillatory function in the mutuality of associations from past life experiences, all individuals do carry the imprint of their consciousness from two opposite polarities, positive to negative, regenerating their frequency and harmonic affiliations from life to life, thus establishing the bias of one's consciousness.

"Therefore, man is not physical even though he appears to be a physical representation in an atomic form. He is only reflecting the terminus of his positive polarity into an atomically-constituted anatomy. And that terminus is the physical body of which the brain is an organ for training and conducting energy as information sine waves.

"The master computer therefore, contains within it the negative polarity, and the positive polarity of itself. That is why it is the master computer; it is the whole note of man. Whereas, what one sees through the physical eyes is the half note of man. One can expand his vision through the accumulation of evidence, verifying the existence of the electronic body, which itself is another link within the interdimensional nature of another dimension of infinity.

"The electronic computer contains the entire map of man's life cycle, a fourth dimensional DNA, in three circuits, the superconscious and the mental conscious represent the positive polarity, while the subconscious represents the negative polarity demodulated into the atomic field structure of a physical anatomy. The physical dynamics of third or material dimensions are an important factor for the experiential growth and development of Homo sapiens. Yet, the psychic or paranormal factors of consciousness are of the greater importance in the individual's spiritual development to selectively regenerate the logic and reason of his progressive evolution.

"The physical anatomy is created at different intervals of time and space. In this respect, these different intervals are indicative that the spirit that activated such physical anatomies or terminals was living in an adjoining dimension within the full nature of its master-computer, or energy body. However as evolution is an alternate - positive to negative regenerative expression, man must experience both polarity functions.

"Simply stated; one reenters another terminal in which

there can be a visible sensing organism. In this manner one gains additional experience from the terrestrial plane or the physical dimension on some planet where the information so gained will be acceptable on a level of understanding for that particular individual. Therefore the process of evolution does contain a double structure or helix in order for man to develop the thought process of logic and reason.

"In order to have access to an atomic world, the terminal through which the individual functions must have component parts that electrically are capable of relaying the signals for man's physical evolution. The evolution of the species Homo sapiens, up to the present is the evolution of the physical anatomy as a sensing device, an instrument, or as a terminal of consciousness.

"The physical terminal is constructed from atomic energy and, as it contains the subconscious derivatives of a lower frequency of infinite intelligence, it is in this respect, an atomic field or the negative polarity of man's electronic computer. The positive polarity or circuit called the superconscious or the higher self, is a product of progressive evolution, and is a development of the inner dynamics or the spiritual nature of that individual.

"In this sense, there is a viable, inner relationship between the dynamic movement of the subconscious and the superconscious energy systems of the individual, whose higher self attains through development knowledge of one's infinite creative intelligence. The higher self is the polarized knowledge of one's creative potential attained from continuous life experiences through the terminal of the physical anatomy. Distilled from innumerable life experiences, the positive polarity, becomes in this respect, the higher frequency print of any individual, and his contact with a higher society of mankind on a fourth dimensional plane.

"Therefore, the evolution of the species man, takes place on two polarities! It takes place on the first hand, in the development of the physical organism, which has now been more or less completed to a great extent - 96% let us say! This physical organism is capable of making access to a higher, intelligent frequency through a terminal, which is a psychic or paranormal development of the individual, who can then express these higher senses.

"Man's spiritual development is the development of higher abilities, a future consciousness that will be relative to his true spiritual identity. That is to say, this circuit that we have called the higher self or superconscious self is a terminal also, but a terminal constructed from a fourth dimensional polarized energy field, distilled from the experiential factors of life-to life living through physical bodies.

"The next development for man will be his capability to operate the complex mechanism of his mind, tuned to a frequency coefficient which can be considered fourth dimensional. Such frequency is not depended upon the atomic coefficients of energy, as it is radiant energy functioning beyond the mathematics of time and space.

"The polarity structure of the mind, and its application by the individual to maintain his equilibrium, is a Science of life, a new mathematics that will be the basis of the new curriculum for the progressive development of ready individuals.

"One cannot discuss the physics of the mind, the higher intelligence of man, without recognizing the impedances which have blocked him from the principles of the logic and reason, the higher constructs of himself. Therefore, to a great extent, in the very act of seeing through the physical eye, one is transposing the past into the present.

"A true relevance as to the cause of any disease must be made from the psychic anatomy and any corrective therapies likewise must be instigated within the psychic body.

"In order to properly analyze and understand the energy spectrums involved in the brain and the psychic anatomy, it is most necessary to make certain differentiations. Any atom radiates an electromagnetic field (EMF). A group of atoms form a molecule radiating an even stronger EMF, one proportionate to the number of atoms and their atomic weights. A large number of molecules form a cell. The human brain contains at least 200 million cells. A subsequent increase in the EMF generated by these atoms is relative to the atom itself, its composite wave forms, etc. The brain, therefore, while it radiates an EMF, radiates information pertinent only to the atom-molecule information contained in the net structure, and very little of any other wave form energies which could be called the emotional content, could be superimposed in this atomically generated EMF.

"Sine waves radiated from a television station contain four different superimposed bands of energy frequencies, sight, sound, picture and synchronization; however, this has been purposely designed to function this way. The wave forms radiated from the human brain are, however, atomically speaking, a heterogeneous assortment of frequencies, all intermixed. The many thousands of these different wave forms are relative only to their particular atom-molecule-cell combination, and due to certain wide differences in frequency bands, are not suitable carriers for the transmission of thought which occurs on a different plane, one more closely fourth-dimensional in its spectral wavelengths.

"Now, in the fourth dimension, there is a certain band of cyclic energies which, like the television carrier, can transmit thought instantaneously, even to remote distances thousands of miles away.

"The brain cells do not and cannot generate electricity, even though electricity in the brain has been measured in certain wave forms by the encephalograph, which is a very sensitive amplifier. Brain cells can alter or change certain wave form characteristics, or they can retain for a millionth of a second a certain wave form facsimile. This retention is called purveyance.

"The net sum and total of all oscillations from the psychic anatomy to the surface exterior life is a process of rectification and purveyance within brain cells, which create the illusion of sight, sound, feeling and different emotions, etc. But in no case, and without any exception, does the brain ever regenerate any electricity. Furthermore, any direction of energies would necessitate some kind of antenna such as the dish-shaped antenna used in radar transmission. Any such radiating mechanism is however, entirely absent in the physical anatomy.

"Physical life is not really physical, because each individual is functioning through a submersible instrument, the atomic body, and he is replicating the present screen of his life, which he had formerly experienced in his past. This is a fundamental principle of the continuity of consciousness explaining the constant relationship of present life as an associated factor from past-life experiences outpictured as one's home, city, male or female bodies, and a host of enumerable familiarities between persons, places and things.

"The sight of a flower will elicit a response of great

transcendency, or it can cause a great sickness to such an individual; all according to the representation of this plant life from previous lifetimes. This speaks about the interdimensional nature of life lived in the present; not necessarily indicating a greater understanding of the spiritual factors of life, but rather, indicating to the individual that he has triggered the encoding signal of his electronic energy body, which now carries the imprint of these traumatic experiences of his life.

"The imprint of positive and negative traumatic life experiences are oscillating within the subconscious circuit of man, an electronic imprint formed from myriads of life to life experiences. An experience itself, is an association of positive to negative information wave forms carried piggyback as sine waves which are constantly oscillating in a cycular manner. The associated daily involvement of any individual, either reflect feelings of calm and well-being or illicit negative responses associated as physical disease and breakdown in mental functions. Such mental and physical reactions may be as paranormal experiences that serve to disassociate the individual from his present physical environment to give him cause for reflection about the state of his mental facilities.

"It is in these examples that the polarity relationship of any individual is shown as the movement of thought within the energy flux fields that serve to change the fabric of one's reaction as the sum and substance of his psychic and physical environment. This principle of frequency and harmonics is then an expressed thought which contains a formed electronic sine wave oscillating within the bias that has been recorded upon it from positive or negative life experience.

"Each individual, therefore, is inputting from his own

computer tapes stored information that contains the bias of his reactions to persons, places and things in his present environment. This is the frequency principle of energy, that once inputted, this energy regenerates from life to life and moment to moment, as harmonics of the initial experience explaining the reality of negative or positive experience affecting any individual in his present life. One is necessarily a victim of the established information that has been formed from the past, which is now on the basis of familiarity, whether positive or negative, carried through this harmonic principle. All factors of life are then really psychic and a paranormal reflection of the continuous movement of consciousness from two disparate reference points held together on the similarity of such experience.

"The act of perception is a state of consciousness, which an individual develops for himself, in the recognition of the alternate nature of the irradiating, electromagnetic flux field which interpenetrates all matter. All individuals have had prior life experiences that set in motion the frequency which could be radiated into the cortex, and thereby, in a nanosecond, the signals carried along the energy highway from the superconscious to the mental conscious and the subconscious connected to the four physical brain lobes completes a full cycle. This cyclic energy matrix is consciousness. It is a result of any and all reactions to the atomic environment and the electronic nature of the psychic anatomy containing past life experiences. Because in all respects, we can see that the picture of life is basically an illusion similar to the pictures on the phosphorus coated screen of the television receiver. In this way all attendant reactions are the completion of a complex movement of energies as sine waves carrying a signal of particular life experiences.

Schematic of the Psychic Anatomy

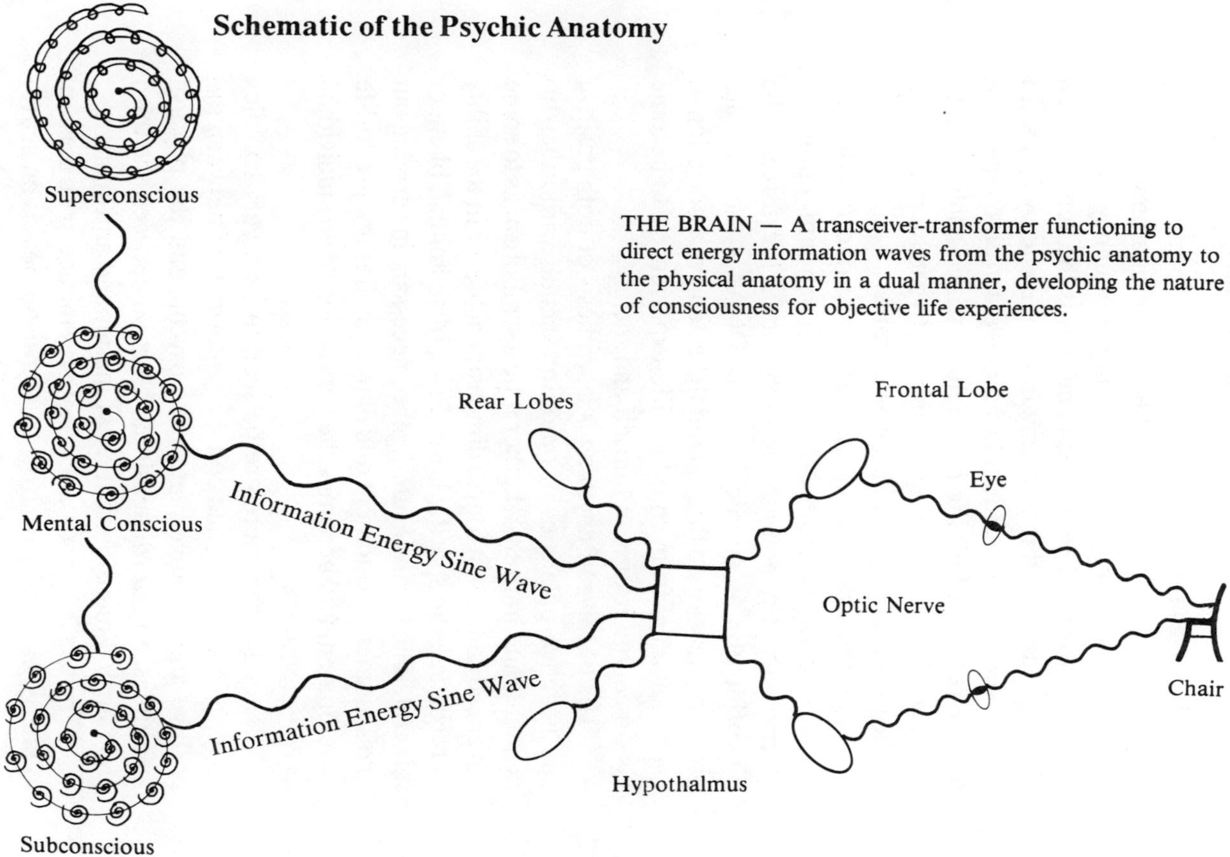

THE BRAIN — A transceiver-transformer functioning to direct energy information waves from the psychic anatomy to the physical anatomy in a dual manner, developing the nature of consciousness for objective life experiences.

"Man lives through the access of two terminals existing in alternative dimensions. The psychic anatomy is electronic in nature and functions on high frequency sinusoidal or carrier waves. The physical anatomy is the demodulated picture of the psychic anatomy in atomic energy form. In other words, the individual has made electronic tapes that are the records of his life experiences, which enable him to conceive the nature of physical life relative to his perception of the present factors in his environment.

"The physical anatomy, a configuration of cell structures assembled in an atomic form, is dependent upon the electromagnetic, non-atomic body for its life force. The blueprint of this atomic form, its DNA, is the software program, developed throughout the extensive association of each individual from countless life-experiences, to become his conceptualization of himself and society.

"To conceive of anything is a synthesis of both positive and negative polarities in one's interaction with other individuals and himself, whether it be a symbolism as to some advanced concept of fourth dimensional life, or in the ability to recognize the specific lower form, of an animal, plant, or mineral. The intrinsic intelligence necessary in carrying out one's moment to moment and daily activities is a life-to-life development and becomes the perception of any individual as consciousness.

"Thought can therefore be seen to be a very complex factor of connecting and collating information between the positive and negative plate - the psychic and the physical anatomies, an interdimensional function between the third and fourth dimensional states of matter and anti-matter.

"The fact is that so-called death is a mirage or an illusion, as energy cannot be destroyed. Consciousness is an activation of thought, a transmission from the two opposite polarities, making possible the physical basis for life, as the negative polarity the physical consciousness makes a 180 degree change in its polarity at the end of the physical cycle, so-called death. Then the same individual continues his consciousness 'on the other side of life', or the astral dimension in which he is expressing from a radiant energy body the electronic or psychic anatomy.

"Death is a contact with some negative past life experience. Man will realize that death is not a conclusion of his life but that he is simply completing the negative end of the polarity principle, within the interdimensional dynamics of his life cycle. For he will continue his life within a few moments, hours, or days, in another environment - one which will seem quite familiar to him! In fact, it will seem as if he had simply been asleep and had just awakened!

"Psychokinesis, life after death, life in other worlds, reincarnation, deja vu, out-of-body experiences, telepathy, etc., will become an understanding of human nature, a factor of greater relationship than can be seen or conceived from the terminal access of the physical anatomy.

"Man lives in two bodies. He lives in two dimensions. He shuttles, to all intents and purposes, from the fourth dimension to the third dimension, the atomic world. On the third dimension he gains important experiences and then returns to an astral planet with an expanded state of consciousness, which of course, is dependent upon many factors in his learning processes of life in which he becomes more in-tune with the interdimensional reality of himself. This back and forth movement is called life. Death becomes another life that is lived in an adjoining fourth dimension!

"Many people call the astral or fourth dimension the

heaven worlds or spiritual dimensions. But that too, is a development in the learning processes of man when he learns to function solely from the mental status of his expanding knowledge of the energy principles. A coherent scientific understanding, using fourth dimensional or paranormal coefficients, can explain the negative polarity of life, time and space, and all aspects of life which are the so-called physical evidence of mass.

"Mass, the energy so associated with the atomic world, is the basic means of advancing the research of each individual into the inner nature of himself. The Eye of the Infinite lies in the individual's mind, and that mind must therefore be the investigative means through which the Infinite nature of consciousness will be discovered

"The dynamics of the mind contains the basis for an expanding consciousness. Consciousness can also reflect expressions of a dilation, or pulls from an individual's past indicating evidences of recriminations, guilts, fears, and other insufficiencies in life. Consciousness also contains a recognition of the reality of the dynamics of the Infinite, a continuous development of intelligence, whereby the negative imbalances of life experiences can be cancelled to attain to the transcended nature of a higher consciousness.

"Emotionalisms, as such, are not a factor of such a higher consciousness. The individual can function without the pressure of a negative subconscious that has grown out of proportion, in the orderly process of man's thoughts as it caresses the Infinite in the recognition of the deeper significance of the continuity of life to its progressive nature in every act, in every incidence of one thought process.

"That realization will move any individual forward when he has that ability to be 'out-of-tune' with the past and to be 'in tune' with the present. Because the present contains and is the balance of this oscillating movement of the past and future. The present can contain the past in such a balanced form where progressive development can be the higher purpose of the individual.

"A let-down in any aspect in life experience, a change in frequency, is always attended to, and reflected from some past association that is now making its entrance into the brain structure of man, which is being rectified and considered to be the nature of his reality!

"Therefore, it would seem only plausible, logical and reasonable that each and every individual would wish to know more about the incidence that has made possible his sickness and to learn the means by which he can correct this sickness. This physics of consciousness is of prime importance to both scientist and layman because these principles are evolutionary factors in the development of the species man. Man, as a product of evolution is striving to become a facsimile of the Infinite through life-to-life experiences to learn of the 'real' nature of his spiritual or energy self!

Chapter 20

THE RENAISSANCE OF LOGIC AND REASON

"Any discussion relevant to information having to do with the large order of things, such as the study of planets and stellar objects, or the study of the microcosmic world does not remove the need to resolve man's basic problems that are factors of daily life. The greater perspective attained in the understanding of the mundane, day-to-day life associations that take up the majority of people's lives and their larger meaning as an interdimensional factor of one's self, provides the intelligence that can resolve the basic pressures of atomic life.

"The study of the mind thereby, is not a study that will enable any individual to escape from the countless interweaving of the webs that he has continued, which is the result of his present life position, as it is reproduced on the photographic plate of his individual consciousness. Yet, it will, in every respect, provide a larger parameter for the individual to institute a new program in his life, gained from a more intelligent perspective as to the evolutionary design of his life, and will provide for such individual a balance in the most minute factors of his daily life.

"Present realizations taking place in the study of the atom, indicates that each subatomic particle follows a basic pathway within the atom's field force. This pathway, associated with the quantum nature of the atom presently being viewed from its negative polarity, will eventually reunite any particle to its positive self as an energy quotient of Infinite Intelligence.

"A particle, no matter that it is a minute elementary

form of energy or is an atomic mass called the physical anatomy, with a spiritual anatomy integrated within it, is still a picture of energy. However, the 'real' person has not been really outpictured and made distinct in form and substance.

"Hence, it can be seen that each individual's mundane life is but a frozen picture in the movement of the total organism. However, with each of these telltale movements of particles, scientists are attempting to understand the total organism of the atom from which this elementary particle has been delineated.

"The total organism of man, although not yet recognized by those who are still developing this image upon the plate of their life, has been achieved by other human beings - Homo sapiens. These individuals have advanced to the next step in their evolutionary pathway of the vortexal sinusoidal expression of Infinity. The understanding of the spiritual cell of an individual, as it is expressed in his present mundane life, third dimensionally speaking, will affect every factor which is correspondingly regenerated in his personal life. In this respect, it can be seen that any individual who develops an awareness of the spiritual factors of life does have a clear, more balanced, and integrated life, even when it is being lived in the elementary particle factor of the third dimension itself!

"In this respect, the third dimensional nature of life, which is the negative polarity of the larger stream of the protonic elements, is the experiencing of the finite particles of life which amplifies the dimension of one's infinite nature.

"The particles within the isotopic field of an atomic configuration are the conclusive statement that life is the experiencing of the many finite relationships within a terrestrial dimension, functioning in the latitudes of light on the atomic mass scale of the elementary tables of the 92 natural elements. Man learns of the reality of the atom through his constant interrelationship within the 'particle' forms of life experience, which have been developed through many hundreds of thousands of lifetimes.

"There is, in fact, energy inundating each individual at every moment in his life. There are bombardments of energy particles from the protonic mass of the sun, and it's electromagnetic field which surrounds each planet! The electromagnetic field surrounding each individual, is being transmitted on frequencies of countless forms to other individual, and at the same moment, this electromagnetic force is interacting between each and every individual throughout his planet.

"Man lives in a society of hundreds of millions of peoples. He is therefore cutting through electromagnetic lines of force that are the accumulations of the life experiences, containing very specific information of other indi-viduals. Each individual is therefore receiving a continuous bombardment of electronic information that is the sum and total of his input and the output of those individuals with whom he is interacting

"Man is receiving the impetus from every individual in his family. As the child grows into adulthood, he emerges into a larger family, the school, society, to eventually emerge into the total organism of the planet, which is the accumulated evidence of the past-life experience of hundreds of millions of peoples who are associated from previous lifetimes.

"So again, we point out that the evidence of living in life-to-life situations, is the learning of the particle nature of the interdimensional cosmos. In this way, each individual is essentially a scientist researcher into his own design, a design that was master-planned by Infinite Intelligence. Each individual has therefore the prerogative, the right, the privilege, and indeed his free will, to determine what he will believe and accept as to the nature of his own life pattern.

"One can appreciate how each and every life becomes an enlargement of his photographic plate, to become the developing consciousness of the individual in each and every lifetime to illuminate the larger design of life. This becomes the illuminated man!

"One can see then, that the Infinite is an infinitely infinite proposition, whereby one life, a million lives, or a hundred million billion lives would not be sufficient to outline the entire parameter, associated with Infinity. A clearer picture of the Infinite would be a sinusoidal expression expanding and contracting in cycular time.

"The Infinite is in constant expansion; therefore, it cannot be understood in one lifetime. In some respects, it is similar to what we have demonstrated in freezing the moving track of an electron to determine the nature of its mass, spin, etc. In fact, the electron is only the product of a continuous frame of a picture, always moving from its left to its right polarity, and is the result of a parallax of energy constituents.

"In this respect, man is very much the same, in that he is a moving particle so associated within the larger matrix of infinity. The photograph of a man on the third dimensional plate is the image reflected from his atomic mass body. If one were to photograph such an individual in his fourth dimensional state, one would view this individual in an entirely different resolution. Such a photographic instrument would reveal the electronic energy body as the total design of such an individual.

"The scientist would exclaim, "Ah, this is the evidence

of the unified field of energy!" It would be a great demonstration to physicists for this is their desire and ambition, to know the answer to the grand unified field equation in which all four factors of energy: gravitation, the electromagnetic, and the weak and strong energies, have been unified within one basic principle of energy!

"That basic principle can only be understood when the spiritual factors of life are understood to be the unified nature of consciousness. If one could photograph the electronic field, radiating from the corona of the physical anatomy of Ruth Norman, they would have evidence of an advanced intelligence who, evolutionary-wise, reflects the accumulated evidence of more than nineteen million years of development!

"As we were saying, those human beings who have graduated from the particle field of terrestrial life and who are capable, mentally-wise, to breathe the higher frequency of an aerated dimension, can function in an electronic body that contains a higher oscillation. That is to say, it has the capability of inputting and outputting; functioning beyond the understanding of present third dimensional electronic science! Such advanced individuals who have learned to fabricate an instrument of pure radiant energy, which functions in a higher frequency field, are capable of recreating a lower frequency instrument, since the master instrument contains all of the potentials of its subdivisions.

"This master instrument is the extended intelligence of the Cosmic Visionary Ruth Norman, expressing the higher dimensional consciousness of Uriel. Such an Intelligence contains the joined factors of Infinity, polarized through and by the Higher Superconsciousness. Thus Ruth Norman is the example of the interdimensional properties of the mind factored as a unified field, an electronic system not limited by dimensional atomic equations.

"The higher equation of this advanced intelligence is a Universal Radiant Infinite Eternal Light! In addition, individuals who have developed themselves mentally are a functioning part of a Brotherhood that is advancing the spiritual factors of humanity throughout terrestrial and astral dimensions.

"That developed electronic body is the superconscious, the architectural rendering of all associated knowledge attained in the understanding the of interdimensional nature of life. Therefore, the superconscious is, in every respect, a product of the infinite creative intelligence!

"Hence, such an intelligence that is now completely unified with Infinite Intelligence is correspondingly an Infinite Intelligence, because it is polarized in its oscillation. There is a great deal more that can be said and will be said in further books of the higher physics of this interdimensional Science. Such a higher physics is a proper study for individuals who have reached that point in their understanding of these scientific principles where they are no longer separated in their understanding of the infinite properties of themselves. Moreover, in this respect, they are no longer third dimensional scientists.

"Hence, the proper study of man is man himself! The study of man will reveal the radiant energy field which is the corresponding proof of his electronic anatomy. This radiant electronic field, which functions on a fourth dimensional basis, can be seen to some extent, with the higher physics involved with the Tesla coil. It is a valid statement that man is an energy system. Moreover, the energy system that he accepts as his basic tenet for life in the day-to-day activities on earth, is a lower factor of the interdimensional energy field of life, and only the design and appearance of a physical anatomy.

"Yes, each individual who is involved in the science of life has moved through his energy sea from the fourth dimension into the third dimension. They are again adding to the sum and substance of the knowledge that has been gained by them in their evolutionary movement. To this extent, with respect to the planet earth, the difficulty in bringing the reality of the fourth dimension or of the interdimensional nature of life to mankind was because of the breakdown in the corresponding positive movement of these electronic signals, which function within each individual, forming the reality of themselves.

"This breakdown has resulted in a short circuit within the instrumentation of man's consciousness, a knowledge of himself as an electronic force. As a whole, in the history of man, there has been a deactivation of the forward movement in the resolution of this understanding of man as a participle of the Cosmic Mind or Infinite Intelligence. The problem really is that those involved in the pursuit of truth have forgotten that they are themselves a part of the evolutionary process of life, and are inputting into the instruments of their present means of research, such as the particle accelerators, etc., and are continuing in this manner, due to their ignorance of life as an energy factor of an interdimensional design.

"Each individual contains a basic lens which is capable of refracting information as a picture, revealing the nature of the atom and stellar objects in the galaxies. Therefore, according to the capabilities of this lens, as with an optical telescope, he will refract that information , to be inputted into his mind.

"Since the majority of earth people have a history associated with breakdowns in the instrument of their mind, their lens is spotted. Therefore, what they do see has been negatively refracted, although they believe they are seeing clearly, they are *deflecting* the real picture being projected within the great sea of energy!

"The mind therefore, as it is being diagrammed, is an electronic instrument that has been rectified by individuals who do not posses a dirty lens, and have not short-circuited their electronic body. Such individuals are, in this respect, capable of delineating a clearer picture of the electromagnetic field which is the property of all human beings! An individual human being who is an earth person, is seeing through a dense overlay of cloud forms, and dimly perceives the bright sun that is the motivating factor for his life and the entire physical terrain! This cloud is the negative accumulation of life experiences and is the content of the subconscious circuit of the electromagnetic field that is man's mind.

"This cloud, for certain people, is thinner; they can therefore, see what lies on the other side. With some, it is denser; with others it is so dense they cannot make out anything other than the nature of their physical body and the terrain in which it is functioning. Those individuals capable of seeing through the cloud cover are Cosmic Visionaries. They are 'clairvoyant'. There are no clouds covering their field of vision. Ruth Norman and others of such development, clearly are not earth people, because they are not clouded with negative energies. This is Consciousness that is, in every respect, associated with the name 'spiritual'!

"Such an individual who has a spiritual Consciousness, is one who can see the state of life of man in his true bearing. He (or she) is able to see the past, the present, and the future within an integrated pattern. The past, as emotional life experiences, has been balanced with knowledge of the alternate stations of life, from its non-atomic to its atomic associations.

"Cosmic Visionaries are, in fact, aware of, and completely in tune with the higher frequency worlds on which people who have gained this development are presently living in tune with the Infinite!

"Yes, the science that we are explaining to you is an interdimensional science for the future of man. It is a science which must be learned by all peoples in their evolutionary development if they wish to experience the consciousness of an expanding Infinite. Then their pulse of life will radiate to provide them with the capability to express these principles to the millions of people who are likewise striving for the understanding of their spiritual nature.

"These advanced individuals are Brothers, in the respect that they are balanced in their awareness of Infinite Intelligence. They are capable, with their polished lens, to project the nature of their Consciousness to all people in all stations of life, while functioning on lower frequency dimensions.

"This means that any intelligent, concerned individual will, by the very logic and reason of these interdimensional principles of physics, recognize the existence of other human beings who have advanced to higher expressions of life and who live on non-atomic planets in worlds where man is expressing as a polarity to each and the other.

"There are other individuals who have reincarnated without an understanding of their association to the interdimensional cosmos. They have accumulated information and gained a great deal of material knowledge while living on an earth planet. Instead of distilling this knowledge to gain an understanding of the Infinite factors of themselves, they further compacted this information and applied it in terms of the atomic factors of life.

"Such individuals have been emperors, kings, queens, as well as scientists and other individuals who have been associated in the descending nature of man's evolution. Great civilizations have seen the evidence of higher principles of life that had been taught to mankind in various civilizations. These ascending spirals which have supported man's positive nature have naturally been recorded in the histories of man, in the Golden Age of Greece, of Egypt, and of Rome and in other civilizations on Earth.

"However, there has been alternately the descending spiral! There have been rampages of powerful peoples who have cut a swath through the planet and destroyed all who have opposed the advance of their armies; the Genghis Khan, Alexander the Great, Napoleon, Hitler and others who, in their concern for self were unconcerned with the spiritual, evolutionary design of life.

"There is, in this respect, in the evidence of the past of man, an opposing movement of this great energy Force, which could be diagrammatically conceived of as expressing the opposite pole of the sine wave principle of life. Therefore, man at the present time, is posed with a great responsibility in his attempt to learn of the spiritual nature and reality of self. He must convert the negative energies which he has inputted into his electronic energy body, energies which surrounded him like a dense cloud cover. The accumulated life experiences of man as a whole are presently being outputted through his grand computer into each individual to become his thoughts and are the nature of his present consciousness.

"This then, is a time of great change! It is the *Spiritual Renaissance of Man.* A greater humanity who, as Brothers on higher frequency worlds, is working to balance the negative polarity of earth by inputting greater positive powers through the knowledge of the interdimensional principles of life!

"These interdimensional principles are lived by Cosmic Visionaries; individuals concerned only with advancing all factors of the Infinite in its progressive development with Homo sapiens. Such Brothers of Light are Unariuns because they are expressing the interdimensional principles of life by functioning to project their Infinite Intelligence. It is the power of this expression of Spirit that is the balancing factor at the present time on earth.

"The purpose of this book is to explain in the language of physics and electronics, knowledge that can be applied by scientists in their respective disciplines: mathematicians, physicists, geologists, neuropsychiatrists, anthropologists, cosmologists, biologists etc. There have been periods throughout man's history where the interdimensional physics of life have been applied with ignorance to technology and has caused a distortion to earth life. The basic proof of this is the explosion of the atomic bomb and the development of the hydrogen bomb.

"As Robert Oppenheimer stated when he witnessed this explosion, 'Death has overcome us!' This indicated his realization of the misalignment of the energy of the Infinite flowing into the third dimension, energies out-of-phase with its fourth dimensional intelligence. This catastrophic explosion is a disease of ignorance which has enveloped the global society of earth. As life is a continuity, this has been the history of man from previous civilizations, and is a factor with each individual who is pursuing science.

"Clearly then, an interdimensional understanding of physics does illustrate the dynamics of the mind; and the dynamics of the mind of man clearly indicates his association in the panoply of life as an ongoing movement of the Infinite. As man is processing his own small energy field to rectify it,to polish his lens, so will it be in tune with a greater intelligence, which is the expressionary factor of Infinite Intelligence.

"This, then, is the basis for the present development in the 20th century, for the New Renaissance of Man as a Spiritual Being. Man will express the nature of himself as an intelligent particle of the greater whole. The greater the attunement to the progressive movement into the wider expanding nature of Infinity, the greater will become each individual, the planet, and the galaxy as a whole.

"Knowledge, of course, is power, but knowledge improperly used is a great negative explosive force! Knowledge that provides wisdom is a positive power, and which integrates each and every atom that is so expressed within the individual man in his third dimensional nature! This integration on the third dimensional side of life is correspondingly integrated into his fourth dimensional body. The higher atom develops from his recognition of the factors of his spiritual nature to then become his photographic Image.

"This plate is man's Consciousness which will be attained with the understanding of the interdimensional principles of the corresponding movement of man from an atomic to a non-atomic body. He will then be expressing as a *Brother of the Light.* He will be expressing from a Light body, fashioned from the *Radiant Energies of the Infinite Creative Intelligence* without the necessity for a secondary body of bone, skin and flesh!

"These integrated principles of life resolve once and for all time the understanding of the past, present and future, integrated with the understanding of the sine wave principle of life. Energy is the prime factor that lives microscopically and macroscopically within man, expressed according to his knowledge of it, expanding and enriching the life of each individual.

"This then is the time of man's *awakening to the truth of his origin and purpose of life!*

Index

Advanced Beings 8,12,67,164,171
as Brothers 93,197,324-332
have communicated 188,211
unified consciousness of
Akhenaton 249,294
Alexander the Great 329
Alfven, Hannes 60
Alignment 170-171,174
misalignment 260
Alternate movement of energy 34,164
Amnesia 86-87,245,253,259
Amoeba 119
Analogy
earth, a classroom 51-54
man on the River of Life 19-23
of the tevision receiver 6,72,78,100
152,264,312-313
Anaxagoras 186
Anaximander 186
Andromeda galaxy 63,126-128,139
Answers 22,28,36,146,227
Antimatter 109,117
Aristotle 186,233-235
Art 3-4
Ascendent spiral 245
Astral body (see electronic body) 297
Astral flight 114
Astrology 136-138
Astronomers 103
Astrophysicists 144,215
Atlantis 56,86-87,136,246-247,251
Atom 27,79,114,171,307
a television picture 6,72
combination 276
formation 111,114-115,139,217
helium 131-132
hydrogen 123,125,127,131,139
intelligence of
interior of 15-16,19-23,214
nuclear force of 115
parallaxes 115
regenerative power of 223,303
unified state of consciousness of 15
Atomic bomb 87-88,102-103,330
Atomic fission 127
Atomic particles 94,214
Atomic structures, origin of 273
Attitudes 190
of the mind 165
Attunement 311
to the electromagnetic spectrum

Bacon, Sir Francis 16
Bar magnet 124,165
Barrier of light 50
Base plane frequency 6,94
Beginnings 57,222,224
of man 71,192,229,292
of stars 60
Beingness of man 241,
Bible 222
Big bang theory 16,45,60-62,101,105
Birth 98-99,101,230-231,299,306
Blocks to realization 36,188,262,312
rectification of 265
Blueprint
of the mind 286
of the physical anatomy 286
for the combinations of atoms 276
Bohm, David 22,27
Bohr, Neils 17,77,203,206-207,211,213
Boltzman, Ludwig 89
Brain 193,201,280
a demodulation of mind 98,289-292
298
a terminal of the Master Computer
98,199,308
a transceiver 98,291
the negative polarity 289
true (see electronic body, mind,
psychic anatomy)
Brain waves
measurement of 134

INDEX

Broglie, Lewis de 204,213
Brothers 90-93,104-107,197,324-332
Buddha 169,294,295

Calculus 21,41
Cancers 264
Carrier wave 35,219,292
Causal force 8,157
Causality of life 283
Cause & effect 61,162,168,173,248
306
is a continuity 177
Cellini, Benvenuto 249
Cell structures 43
and DNA 272,285,298
development of 119,230,273-275
electronic nature of 285
growth of 185
power of 280
regeneration of 275-276,280,303
Channel for consciousness 5,8,14,175
265
Chemical dependence 103
Chromosomes 193
Civilizations
previous 87-88,245,252-253
rise & decline of 255,268
Clairvoyance 186
Cognitive ability 296
Complimentarism, theory of 17
Computer science 257
Conscious mind 98
Consciousness 19,31,40,88,176,231
237,306-319
a closed circuit 291
a product of an imprint 310-314
and exchange of intelligence 296
and perception 293
and thought patterns (see thoughts)
as a channel 5,8,14,175,265
continuity of 17,56,82,232,253,290
Cosmic 183,199,307
dichotomy of 81,235
duality of 205
dynamics of mind 58,191318
expanded 91
higher 147,211
infinite 178
interdimensional movement of
206-213
isotopic nature 219
mental 263,315
transcended states of 46-47,52,242
267
true 78
spiritual consciousness 167,327
totality of 208,219
unified state of 11,82,209,252,324
Continuity
between 3rd & 4th dimension 219
279
cause & effect 177
evidence of 282,286
of consciousness 7,56,82,206,232
253,290,314
of actions 87
of energy 48,82
of life 153,242,244
of life after death 98,195,222,231
251
of previous life times 151,190
253-271
principle of life 36,177,220,260
Copernicus 21,149,234,244
Corridor of time (Egyptian pyramid)
136
Cosmic
centrifuge 63-66,110,123-124,128
134,141
Consciousness 183,307
egg 100,104,183,186
hysteresis 110-111,126,133
Intelligence 185
Mind 2-3,48,69,106,183,186
Visionary 327-328,330
Cosmological Constant 67,68-69,100
149
Cosmologists 104,144,215
Cosmology 237
Creation
of man 229
of the universe 63,66,150
Creationism 240-241
Creative force (see also mind) 8,104
230
Cycle of the recessional 222
Sixth 12,46
Darwinism 240
De Broglie, Louis 206
Death 73,82,97,107,166,168,222,226
230-231,270,279-280,299,307
an illusion 316
life after 98,195,251,317-318
memory of 300
Descartes, Rene 31,43,59,147
Demodulation 79,244
Descendent spiral of life 13,151,329

INDEX

Deja vu 317
Dichotomy
between mind & body 118,287
of consciousness 81,235
Dimensions 39,45
higher 81,211
relationship between 32,279
Dirac, Paul 29,204,206
Disease 200,254,260,262,306,312
contagious 264
Doppler effect 60,61-64,78
Double helix 272-273,285
DNA code 193,272,280,285,309-310
316
Dynamic energy envelope 191

$E=MC^2$ 84-85,279,290
Earth, planet 74
Earth life
a classroom 51-54
Eddington, Sir Arthur 204
Edison, Thomas 32
Ego 117,221
deflations 198
developments 190
higher 170
structure 44
Einstein, Albert 16-17,22,29,32-34,77
84-85,89,93,204,206,211,213,249
Special theory of relativity 93
Theory of relativity 63-64,
El Greco (Domenico Theotocopuli) 2,9
Electrochemical reactions 135
Electromagnetic body (see Electronic body)
Electromagnetic fields 115,192,235
243,286,312
measurement of 134
of planets 223
Electromagnetic force
within man & universe 18,322
Electromagnetic lines of force 18,28,68
94,112,123-126,144,213
Electromagnetic spectrum fields 29
67,134,165,284
Electron(s) 83,91,253
Electron microscope 205,304
Electron wave forms 272
experiences as 258,263,314
polarized 269
rectifying 265
regeneration of 172
thought forms 266-270
Electronic body 11,47,89,185,197,217
329231,325
a transmitter 81,251,291
astral body 293,301,303
the master/grand computer 178,199,
256,292,306-310,
three circuits of 290-299,309-311
Emotionalisms 174,319
Encephalograph 134
Energy 33,187,215
a reflection of the Infinite 187
alternate movement of 34,75,96,164
255
coexistent with other fields 46
continuity of 17,48
conversion 134
cycular in form 85,96,111,138
degenerating movement of 13,151
four forces of 70,75,96,113,324
harmonic regeneration 63,112,135
higher frequency 23-24
intelligence of 225,296
interdimensional movement 73-74
95,122-129,130-142,264
isotopic nature 217
measurement of 79
misuse of 179
moving on subatomic levels 16
mutation of 262-266
oscillation 164-165,210,218,256
209-210,218,299,313
out-of-phase 104
plasma sea of 70,122,138
polarity reversal 96,110-111
psychokinesis of 14,306
rectifying the wave form 265
regeneration of 127-128,172-173
vortexal nature of 22,94
unified field 236,252,324-325
wave forms 262-266,308
Energy body 85,217
Energy of man
is divisible 40
Enigma of creation 150
Entities, nonatomic (see also obsessions) 268
Equilibrium 74
Eros 197
Evolution 33,147,214-216
basis of 24-25
nature of 167
physics of 239
progressive 201,308,329
purpose of 232

INDEX

regressive 329
theory of 240
Evolutionary
cycles 222
design of life 41,185
development of mankind 105,290
process 150
purpose 232
Expanding universe 26,244
Experiences
imprinted as electronic wave forms 258-266,308-314
negative 166,329
out-of-body 195,306,317
paranormal 205,281,314
polarized 39,98
previous life 9,36,46,200,258,295,315
transcendent 46-47,242,267
trial & error 180
traumatic 190,263,302,190,263 302,314
Extraterrestrial intelligence 28,76,105

Father 2,48
Feedback 7
Fermat's Last Theorem 25,39-40,18 283
Fission, atomic 127
Fetus 289,295
Five senses 43,48,193
Finite dimension 299
Fountainhead 2,48,73,
Four forces of energy 70,75,96,113 324
Fourth dimension 49-50,146
non-material dimension 78
positive polarity 89,313
transmitting field 160-161
unified field 96,103,213,252,
Forth dimensional
physics 152-155,224,287-288
scientists 154-155,187-188
Franklin, Benjamin 165
Frequency(ies) 6,41,209-212,311
fourth dimensional 111
higher 23
of an induvidual 211
Fusion 62,127

Galaxies 223
development of 71
not receding 64-65

Galileo 214
Gamma rays 60
Gauss, Johann, 204
Genetic code 193,272-273
Genetic development 101
Genghis Khan 329
God 2,214
Grand Design of Life 210
Grand Unified Theory 144,157,252

Harmonic affiliations 297-298,302,309 315
Harmonic regeneration 63,112,135
Harvey, William 193
Healing 266
Heat 64,113,132-133
Heisenberg, Werner 29,203,206-207 213
Helium
formation 131-132
Helmholtz, Hermann 204
Hermes 169,186,249
Heterogeneous nature of life
Higher
carrier wave 158,197,292
civilization 160
Consciousness 147,211
Creative force 99,104
dimensions 5,205,211-213,224
frequencies 23
Intelligence 46,49
Intelligent Forces (see advanced Beings)
Mind 5,
Self 7,26,170,269,310-311
sense structures 5-7,155,281,310-311
transmitter 211
Hindu 180
History of the individual 303
History of man 6-8,36,56,74,169,331
History of science 32,220
Hitler, Adolf 329
Hubble model 61
Hydrogen 123,131-132
universal element 67
Hypnosis 306
Hypothalamus 289
Hypothesis 147,212,239
Hysteresis 110-111,126,133

Immortality 136
Individuality 227

INDEX

Infinite 27,40,228,323
  and finite 27
  consciousness 178
Infinity 116-118
  ascendent spiral of 245
  descendent spiral of 151,329
  is incremental 295
Infinite Intelligence
  a constant expansion 323
  cell of the (see electronic anatomy)
  Concept of Creation 63,66,150
  power of 165
  sparks of the 10,161
  vortex 144,215
Information
  network within man 253,295
  technology 246
Inspiration 12-13,107,237,267
Instrumentation 134,205
  dependence upon 92
  present 65,146,242
Integration 98
  between mind and body 75,118,175
    248
  disintegration 254
Intelligence 225,228
  and consciousness 296
  and perception 293
  Cosmic 185
  extraterrestrial 28,74,105
Interdimensional
  awareness 234
  communication 188,211,216
  knowledge 143
  movement of energy 66,73-74
  movement of life 45,287
  physics 184,187,242-243
  vortex 160,215
Interpretation of data 81,224
Interrelationship
  centrifugal to centripetal 214-218
  of information 287-295
  mind & body 118,286,296
  subject & object 184,237,239,247
  transmitter & receiver 7,286-287
    291-299
Isochronisms 94,123,135

James, William 262
Jesus of Nazareth 169,294,295

Kant, Immanuel 239
Kepler, Johannes 21,30,43,234
    243-244
Kirlian Effect 285
Knowledge
  is power 331
  of the mechanics of mind 12,265
Krishna 294
Kung Fu 95

Laser beam 116
Leibnitz, Gottfried 43,147
Lemuria 136
Lens of consciousness 186,211
    215-216
Libido of life 160-161
Life 287
  a continuity 8,47,151-153,242-244
    260,277-279,330
  a development 54,287
  an energy system 5,69,325
  as a television receiver 6,254
  continuity after death 98,195,231
    251
  cyclic movement 27,230-232,283
  descendent spiral of 13,151,329
  evolutionary designs of 53
  experiences 98,186,258
  mechanics of 166
  paranormal factors of 5,281,311
  previous associations (see harmonic
    affiliations)
  previous experiences 19,36,46,200
    258,315
  purpose of 99,161,228,332
  Science of 175,311
Light 45,85
  anomaly of 182
  barrier 21,49
  body (see psychic anatomy)
  speed of 17,37,43,46,85,114
Logic & reason 44,169,203,210,282
Looking glass Universe 78,181,183
    185

Macrocosm 39,70,150,159,215
Malfunctions 267
Man 157,165-166,321
  a developmental relationship 40
  a measuring device 69,81,157
  a particle of the Infinite 176,323
  a spark of the Infinite 9,161
  a spiritual being 2-4,41,179,331

INDEX

a transmitter 81-82,97-98,181
an electronic instrument 6,11,260 291
an energy system 7,47
has two homes 32,39,167,317
history of 5-8,36,56,74,169,331
memory of 252,289
reflecting from two polarities 294-298,309
the research project 248,277,322
Man's birthright 9
Man's computer 256,292,306-310
Man's consciousness 19,31,40-41,91 192
a continuity 82,232,290
paranormal 97
Man's development 18,20,67,71,221
origins 192,240-242
present 11,
spiritual 309
Man's lens 186,215
Man's mentality 254
a computer 121,256,292
a sensing device 147,157
Man's mind 42,80-81,100,235
a non-atomic plasma 67
a subject for research 281,287
barriers of 227-230
electromagnetic effect 70
functions in two dimensions 32,78 269,291-305,317
positive polarity 289,293-296
true measuring devices 205
Man's purpose 178
Man's responsibility 13,323,329
Mapping
of life 153
the 3rd dimension 148
of fourth dimension 154
Mass 294
a partial picture 279-280
Master Computer (see electronic body)
Mathematical symbolism 27,92,284
Mathematicians 274
Mathematics
a reflection of one's concept 42
as a symbolic logic 202
Maxwell, James clerk 22

Measuring devices 81-82,99,204
man as 67
spectroscope 60-63
Memory 252,289,29
Meng-tse 295
Mental
blocks 36,48,188
body 68-70,226,289-292,296,298,
breakdown 200,260,306,326-327
consciousness 263,315
disintegration 254
Mental retardation 301
Metaphysics 202-203,224.234,239
Michelangelo, Buonarrioti 1-4,9
Micro energy fields 43
Microcosm 15,39,70,150,159,215
Milton, John 164
Mind 11,37,52,173,175
a 4th dimensional development 286
a bridge 32,281
a computer 19,148,256,292
a mechanism 175,257,260
a positive polarity 289,293-296
an anatomy 51,78
an electromagnetic field 235-237
cosmic 2-3,48,106,183,186
dynamic process of 58,191
electromagnetic effects of 68-70
hypothalamus, the link 289
integration of 70,176
interrelationship with body 286,296
is his software 316
is interdimensional 32,78,155,269
mechanics of 265
physics of 152
the transmitter 81,97-98,181,251,291
Missing link 190,226
Mitosis 9,98,296
Mirror images (see also looking glass universe) 100,208,216
Molecules 276
formation 119,214,273
Motivating force 8,162
Movement of energy 12,
alternate 34,96,164
as thought 153,265,308
degenerating 13
expanding & contracting 70
interdimensional 73-74,95,264
Multiple sclerosis 264
Muscular dystrophy 264
Mutation 262-266
Mysticism 180,203

Napoleon 329
Negatively biased individual 247,300
Negative experiences 166,329
Negative polarity 193,290,329

third dimension 321
Negative polarity 309
New Age of Logic & Reason 25
New mathematics 16,42,45,83
New physics 21-22,38,89,155,182,306
New World Teaching 86
Newton, Sir Issac 21,30,,234,243
Newtonian physics 16,34,37,
Norman Ernest L. 158,189,197,299
mapped the Interdimensional Cosmos 154-155
Norman, Ruth 324,327
Nucleus formation 111-116,293

Objective measurement 78,83-84,87
Objectification 293
Objectivity 239,247-248,304
Obsessional forms 268
Obsessions 262-271,306
Oppenheimer, Robert 16,21,32,330
Orion nebula 60
Osiris 294
Out-of-body experiences 195,306,317
Out of tune 174
Oscillation 164-165,210,218,256,299
between mind and body 75,118,313
demodulated 209-210
negative 165
parallel worlds 118

Parallel universes 109,116,118,141,167
Parallaxes 115
Paranormal 97
abilities 81,85,307
development 311
experience 205,281,314
factors of life 5,281
Parapsychology 194-195,
Particle accelerators 34,94,159,181 238,277,305
Particle physicists 144,215
Past civilizations 86-88,252-253
Past life associations
and the subconscious 295
memory of 252-260
Past life experiences 19,22,24,259-260
rectified 256
Past life regression 306
Past life remembrances 87-88,252-260,
Pauli, Wolfgang 29,203,206,213
Perception 208,293,315
Philosophers 274
Philosopher's stone 186
Physical anatomy/body 2,7-8,48-49,71 77,118,210,228
a receiver 8,97-99
a reflection 289-295,304
a terminal 310
basis for intelligence 277
blueprint of 276,281
breakdown in 200,264
demodulated psychic anatomy 316
evolution of 95
interrelationship with mind 286,296
mutations of 262-266
negative polarity 193
Physical consciousness 167
Physical man 71,313
Physicists 203
Physics
interdimensional 184,187,242-243
of Consciousness 152,156,331
revolution in 29
third dimensional 204
Pierce, Charles 262
Plato 186,233-235,249
Planet(s), an energy structure 126
Planck, Max 15,22,29,32,34,83,93,204 206,213
Planck's constant 23,208
Planck moment 144,
Plasma field 70,122-123
vortical nature 65
Plasma universe 141
principles of 57,72
Polarity(ies) 89,100,107,166-167,190 308
centrifugal to centripetal
negative 193,329
positive 89
principle of 189,191
relationships 124-125,175,315
reversal 96,110-111
Polarization 166,175,269
Political exercise of power 151
Poltergeist 306
Positive polarity 303,309-310
fourth dimension 89
mind 290
Positively biased individual
Possession 306
Power 226
depletion of 177
from alignment 170-171,174
from attunement 165,170-171
Infinite 165

INDEX

of cells 280
permancy 165
regenerative 223,303
Preconditioning 212,277
Previous civilizations 87-88,245,252
Previous lifetimes 9,36,46,200,258,395
315,322
Primary energy structure 272
Principle of polarity 189,191
Principle of cause & effect 61,162,168
173,177,248,306
Progressive evolution 13-14,74,107
171,201,229,308
aborting of 253
and science 38-39,73
Promethium 127
Proton 229,263
Protonic force(s) 22,27,102,143,155
257
Psychic abilities 81,311
Psychic amnesia 8,87,245,253
Psychic anatomy (see also electronic
body, mind) 59,118,184-185,312
a transmitter 97-99,181,251,291,295
and past memories 300
Psychodynamic
process 210
principle 208-209
of the mind 190,194,308
Psychodynamics 235
Psychokinesis 14,195,317
Psychologists 274
Psychosis 264
Purpose
of evolution 232
man's 178
of life 99,228
Purveyance 306,313
Pythagoras 86

Quantum mechanics 34,92-93
Quantum Physics 15-16,
Quasars 60,159
Quarks 159,181,214

Raphael, Sanzio 3,9,
Recognition of the past 252-260
Rectification 306,313
negative wave forms 265
Red shift 60,62-64,78
Regeneration 222-226
harmonic 63,112,135
of cells 276,280,303
of energy 123-128,172-173
of knowledge 224
Reincarnation 279,317,323
Relativity Theory 63-64
Religion 151,202-203,274,306
Remorse 178-179
Research 8,223-224
man is subject 108,184,247
Researcher 23,35-36,152,227,235-236
a factor in the research 31-32,224
a measuring device 81,84
involvement 31,99,277
River of Life 19-23,35

Schrodinger, Irwin 22,29,83,89,93,204
206,208,213
Science 22,23,28,31,49,86,168
202-203248
a continuity 88-89
a roadmap 218-219,250
and progressive evolution 38,49
analytical technique 203
history of 32,220
of healing 266
the highway of evolution 218
Unarius 158,178,195-197,304
Science of Life 75,175,248,312
Scientists 44,49,84-85
fourth dimensional 59
Screens of life 78,152
Self-understanding 150
Senses
higher 5-7,155,281,310-311
physical 43,48,193,299
Shakespeare 164
Siberian crater 109
Sixth Cycle of the Recessional 12,46
Sleep 300
Society 52-54,91-92,269
Socrates 169,249,294,295
Solar energy 62-64
Space/time 42-55,79,220
Spaceships 76,114
Sparks of the Infinite 9,161
Spectrographic analysis 63-65
Spectrums 165,182,284
Speed of light 17,37,43,46,85,114
Spinoza, Benedictus De 59,169,239
243
Spirit 219,228,310
Spiritual
amnesia 87,259

anatomy (see electronic body
mind, psychic anatomy) 185
being 179
consciousness 167,327
development 309,323
dimensions 81,211
identity 311
origin 242
senses 5-7,155,281,311
spirituality 227
Spiritualization 13-14
Standing waves 301-303
Static field 262,266
Subatomic particles 37,77,92,214
320-322
Subatomic physics 91
Subconscious 40,98,232,259,
a subsidiary of the whole 295
builds a physical anatomy 258
functional relationship 263
rectification of 265
Sun 64,111,123-124,127-128,130-135
223
Sunspots 136,140
Stars 123
Super Ego 170
Super nova 60
Superconsciousness 26,170,232,315
the unified field 86,208,219
Superconscious Self 39,174-175,298
311257-259,298
Superforce 100
Swedenborg, Emanuel 239
Symbolic logic 202
Symbolism 18,23,288,314
of the calculus 21,40,92,284

Tai Ching 295
Tao 180-181
Teachers & Masters 88-90
Teaching centers 120
Technology 58,159,221,246
Telepathy 195,306,317
Television 6,72,100,254,312
Tennyson, Alfred 164
Tesla, Nikola 32,177
Tesla coil 325
Theory of complimentarism 17
Theory of relativity 63-64,93
Thinking processes 102
Third dimension
a classroom 51-53,156
a finite attribute of the Infinite 299
a mirror image 100,208-210,216
a negative polarity 321
a video screen of life 78,152
as a spectrum 284
mapping of 148154
Thought forms 266,268,270
Thoughts 153,308,313
rectification 265
Time 41,49
Time & spaee 42-54
demodulations 79
Topological mathematics 78,92,147
Transmitters 81-82,95-98,181,251
291,313
Electronic body 81,251,291
from a higher dimension 28,211-213
Interdimensional 7,291-299
indication of a 101
psychic anatomy 97-99,181,251,291
295
Transendended states of
consciousness 46-47,242
Transcendency 52,267
Trauma 190,263,302
Trial & error 180
Twentieth centry science 248

Unariun Brothers 90-93,104-107,197
218,324-332
Unarius 106,158,160,211-212,275
Academy of Science 197
an acronym 204
an Interdimensional Science 21
library 197
Science 178,195,197,305
Unified consciousness 10,15,80,205
Unified field force 57,80,103,213,216
252
a electronic system 325
of the four forces of energy 75
the superconsciousness 325
Unified field theory 77,93,
Unified field equation 114,212,252
Universe 70,76,185
a functioning mechanism 163
creation of 63,66,150
coexistence between 73
expansion of 101
causation 101,157
mirror & real 107
parallel 109,118,145
plasma 57,70-72
URIEL 324-327

INDEX

Value systems 237
Van Allen radiation belts 113
Venturi tube 185,215
Vinci, Leonardo da 1-4,9,249
Vishnu 294
Visionaries 221,324,327-328,330
Vortical energies 22,94
  Interdimensional 160,215

Way Showers (see Teachers and Masters) 245
Weather patterns 67
Wheeler, John 27
Whole number 17,25-26,40,156
Wigner, Eugene 17,31
Wisdom 186,261

Zen 180
Zoroaster 169